Advances in Analytical and Coordination Chemistry - Applications and Innovations

Edited by Mozaniel Santana de Oliveira, Berta Barta Holló, Muhammad Zafar, Eloisa Helena De Aguiar Andrade and Mirjana M. Radanović

Published in London, United Kingdom

Advances in Analytical and Coordination Chemistry - Applications and Innovations
http://dx.doi.org/10.5772/intechopen.1004667
Edited by Mozaniel Santana de Oliveira, Berta Barta Holló, Muhammad Zafar, Eloisa Helena De Aguiar Andrade and Mirjana M. Radanović

Contributors
Clarence Ng, K. Senthil Kumar, Kareti Srinivasa Rao, Marijana S. Kostić, Mirjana M. Radanović, Paranthaman Subash, Qiao Cao, Ruogu Tang, Sulekha Khute, Tariq Almubarak, Vukadin M. Leovac, Xiangyu Chen

First published in London, United Kingdom, 2025 by IntechOpen
IntechOpen is the global imprint of INTECHOPEN LIMITED, registered in England and Wales, registration number: 11086078, 167-169 Great Portland Street, London, W1W 5PF, United Kingdom

For EU product safety concerns: IN TECH d.o.o., Prolaz Marije Krucifikse Kozulić 3, 51000 Rijeka, Croatia, info@intechopen.com or visit our website at intechopen.com.

British Library Cataloguing-in-Publication Data
A catalogue record for this book is available from the British Library

Advances in Analytical and Coordination Chemistry - Applications and Innovations
Edited by Mozaniel Santana de Oliveira, Berta Barta Holló, Muhammad Zafar, Eloisa Helena De Aguiar Andrade and Mirjana M. Radanović
p. cm.
Print ISBN 978-1-83634-338-7
Online ISBN 978-1-83634-337-0
eBook (PDF) ISBN 978-1-83634-339-4

If disposing of this product, please recycle the paper responsibly.

Meet the editors

Dr. Mozaniel Santana de Oliveira holds a Ph.D. in Food Science and Technology. His expertise includes essential oil extraction from plant matrices using supercritical fluids and conventional methods, focusing on agriculture, pharmacology, and biotechnology applications. He has authored 229 scientific works, including 107 international journal articles, 10 books, and 50 book chapters. Dr. Oliveira is a professor at the Graduate Program in Biological Sciences – Tropical Botany (PPGBOT) from the Universidade Federal Rural da Amazônia and the Museu Paraense Emílio Goeldi (UFRA/MPEG) and a collaborator at the Graduate Program in Pharmaceutical Sciences (PPGCF) and the Graduate Program in Food Science and Technology (PPGCTA) at the Federal University of Pará (UFPA), with extensive editorial experience and an H-index of 23 in Web of Science, 23 in Scopus, and 29 in Google Scholar.

Dr. Berta Barta Holló, an Associate Professor, earned her degree in Biochemistry and completed her doctoral studies in Coordination Chemistry at the Faculty of Sciences at the University of Novi Sad in Serbia. Nowadays, she teaches lectures in several courses from inorganic chemistry and thermal analysis at the same faculty for BSc, MSc, and Ph.D. students. Her research activities focus on the thermal analysis of inorganic compounds and materials, synthesis, physicochemical and thermal characterization of transition metal complexes, MOFs, and coordination polymers with N-heterocyclic ligands or linkers. She is a team member for coordination chemistry at the Faculty of Sciences at the University of Novi Sad. Besides that, she works with several research groups to characterize new coordination complexes.

Dr. Muhammad Zafar completed his Ph.D. in Plant Sciences from the Quaid-i-Azam University in Islamabad, Pakistan in 2011. Currently, he is working as an Associate Professor in the Department of Plant Sciences at Quaid-i-Azam University in Islamabad, Pakistan. He is a renowned scientist in the field of medicinal plants, plant biodiversity, and plant systematics and biodiversity. He founded the first Melissopalynology and traditional medicinal plants labs at Quaid-i-Azam University. Dr. Zafar has over 360 impact factor publications and ISI Web Sciences (citations ±7750, H-index 49), Google Scholar Citation (citations ±14500, H-index 60, i10-index 311) and in diverse fields of Plant Sciences. In addition to this, he is the author of 14 international books and 25 book chapters. He is the editorial board member of various reputed journals. He has supervised 18 Ph.D. and 72 M. Phil. research scholars in the fields of Advanced Plant Systematics and Biodiversity. His contributions have been recognized with several prestigious awards, including the Productive Scientist

Awards by the Pakistan Council for Science and Technology (PCST) (2009–2017), the Tenure Track System (TTS) Awards for Performance-Based Increment (2016–2020), the Gold Medal (Prof. Dr. Zabta Khan Shinwari Award) from the Pakistan Academy of Sciences in Biotechnology (2021), and the Gold Medal from the Pakistan Academy of Sciences in Botany (2011).

Eloisa Helena De Aguiar Andrade holds a degree in Pharmacy from the Federal University of Pará (1980), a degree in Biochemistry from the Federal University of Pará (1982), a master's degree in Chemistry of Natural Products from the Federal University of Pará (1992) and a Ph.D. in Chemistry (2008) from the Federal University of Pará. She currently serves as a Full Researcher II in the Botany Coordination at the Museu Paraense Emílio Goeldi, where she manages the Adolpho Ducke Multiuser Laboratory (LAD). Additionally, she holds the position of Associate Professor I at the Faculty of Chemistry at the Federal University of Pará (UFPA). She is a professor in the Postgraduate Programs in Chemistry at UFPA, the PPG in Biological Sciences - Tropical Botany at the Federal Rural University of the Amazon (UFRA) / Museum Paraense Emílio Goeldi (MPEG), and the Postgraduate Program in Biodiversity Biotechnology within the Bionorte Network. She also coordinates the Pará State Hub of the Postgraduate Program in Biodiversity and Biotechnology (PPGBIONORTE-PA) of the Bionorte Network, having served in this role from 2016 to 2020 and again from 2022 to 2024. She is the author of more than 500 scientific contributions, including articles, communications at events, book chapters and books. She has experience in the area of Chemistry, with emphasis on the Chemistry of Natural Products, working mainly on the following topics: gas chromatography, gas chromatography/mass spectrometry in volatile and fixed (derivatized) chemical constituents, etc.

Dr. Mirjana Radanović, Associate Professor, graduated and finished her master's and doctoral studies at the Department of Chemistry at the Faculty of Sciences, University of Novi Sad. She teaches lectures and laboratory sessions within General and Inorganic Chemistry courses for BSc, MSc, and Ph.D. students. Her scientific interest focuses on syntheses and physicochemical and structural characterization of new transition metal complexes with bioactive Schiff base ligands. Recently, she extended her scientific research to the field of MOFs(metal-organic frameworks) and coordination polymers. She is a part of the research group working on synthesizing and characterizing new porous coordination polymers for CO_2 capture. She is the team leader of the ongoing project.

Contents

Preface

Analytical and coordination chemistry is an area of knowledge that is still expanding, driven by the constant search for innovation and the need for solutions to increasingly complex challenges. This book, *Advances in Analytical and Coordination Chemistry – Applications and Innovations*, was created to bring together some of the most recent discoveries and advances in this field, showing how cutting-edge research transforms theory into practical applications that impact our daily lives.

Organized into three sections, the book covers topics ranging from fundamental chemistry to the most advanced technological applications. The first section, "Schiff Bases and Chelating Agents", explores the versatility of Schiff bases and their metal complexes, highlighting their role in materials science, energy and industrial processes. From compounds derived from adamantane to their use in solar cells and the petroleum industry, the chapters in this section show how these compounds continue to surprise with their usefulness and potential.

The second section, "Innovations in Chromatography, Microfluidics, and Laboratory Diagnostics", focuses on analytical techniques that are revolutionizing the way we study and interpret the world around us. Advanced chromatographic methods for isolating natural molecules with pharmacological potential, as well as the new frontiers of microfluidics, are the topics covered. These chapters reflect the interdisciplinary nature of analytical chemistry and its importance in areas such as health, the environment, and technology. Finally, the third section, "Crystallization", provides a detailed discussion of crystallization processes and their impact on material properties. A specific study of sulfur curing systems and their effect on the vulcanization and mechanical performance of elastomers is a practical example of how controlling these processes can lead to significant improvements in industrial applications. This book is the result of the dedicated work of expert researchers from different parts of the world, each bringing their own expertise and unique perspective. We aim for it to serve as a source of inspiration and knowledge for students, teachers, and industry professionals, providing a comprehensive overview of recent advances and future trends in analytical and coordination chemistry.

We are grateful to all the authors who contributed chapters and shared their research and insights. And, of course, a special thanks to you, the reader, for your interest in this fascinating topic. We hope this book will inform and inspire new ideas and discoveries that will continue to drive this dynamic field forward.

Dr. Mozaniel Santana de Oliveira and Dr. Eloisa Helena de Aguiar Andrade
Museu Paraense Emílio Goeldi,
Belém, Brazil

Dr. Berta Barta Holló and Dr. Mirjana M. Radanovic
University of Novi Sad,
Novi Sad, Serbia

Dr. Muhammad Zafar
Quaid-i-Azam University,
Islamabad, Pakistan

Section 1

Schiff Bases and Chelating Agents

Chapter 1

Complexes of Adamantane-Derived Schiff Bases

Marijana S. Kostić and Vukadin M. Leovac

Abstract

In light of the diverse chemical properties inherent in adamantane and its derivatives, this section shows an overview on adamantane-derived Schiff bases. Adamantane compounds, known for their diverse therapeutic applications, play a crucial role in drug discovery due to their hydrophobic nature, enabling modifications across different drug classes. The rigid cage structure of adamantane shields adjacent functional groups from metabolic cleavage, thereby augmenting drug stability and distribution in blood plasma. These derivatives and their complexes have exhibited the capability to disrupt various enzymes, showcasing a wide array of therapeutic activities, including anti-inflammatory, antiviral, antibacterial, antimicrobial, anticancer, anti-Parkinson, and antidiabetic effects. Highlighting the distinctive chemical attributes of the adamantyl scaffold, the significance of synthesizing and characterizing novel adamantane-derived compounds is underscored. The overview and comparative structural analysis of Schiff bases with adamantane moiety and their complex compounds contribute valuable insights to the realms of chemical synthesis, materials science, and medicinal chemistry.

Keywords: Schiff bases, adamantane, metal complexes, crystal structure, CSD search

1. Introduction

Adamantane, the smallest representative of diamondoids, is a highly symmetrical polycyclic cage molecule with unique properties. Amantadine (**Figure 1**), 1-aminoadamantane, the first adamantane compound used in medicinal chemistry, exhibited potent anti-influenza A activity, initiating the era of adamantane-based drug discovery [1]. This compound has later been recognized for its antibacterial and antifungal properties as well [2]. Adamantane and its derivatives, recognized for their diverse pharmacological effects [3], are frequently employed in drug discovery due to adamantane's hydrophobic nature, modifying properties across drug classes [4–6]. These derivatives have been discovered to disrupt various enzymes, showcasing diverse therapeutic activities such as anti-inflammatory [7], anti-viral [8], and anti-Parkinson [9]. They serve as antimicrobial-anticancer DHFR inhibitors [10–12], and they are also known as antidiabetic agents [13]. The incorporation of the adamantane moiety enhances CNS penetration, making it valuable for targeting CNS drugs [3]. Notably, some derivatives have demonstrated potent anticonvulsant activity, including NMDA-activated channel blockade [14–18]. Additionally, bananins, a class of antiviral compounds featuring a

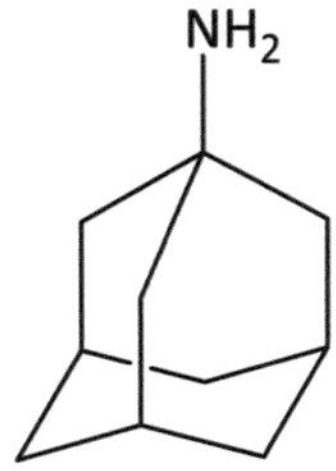

Figure 1.
The structure of the amantadine.

trioxa-adamantane moiety bound to a pyridoxal derivative, exhibit efficacy against both HCoV-OC43 and SARS-CoV-1 [19, 20], while spiroadamantane amine has proven effective against the coronavirus strain 229E [21].

It is interesting to mention that three adamantane-based COFs with varying pore sizes were synthesized using adamantane as a monomer and nitrogen-containing compounds as building blocks through a Schiff base reaction. The inclusion of rigid adamantane enhances the mechanical and thermal stability of the COFs, while the nitrogen-containing building blocks are anticipated to improve adsorption capacity [22].

Various metal complexes of amantadine-based ligands, including platinum(II) and platinum(IV) complexes known for their anticancer efficacy, have been synthesized and documented in the literature. Silver complexes with amantadine and memantine have been reported as potent antibacterial agents against both Gram-positive and Gram-negative strains [13].

The distinctive attributes of the adamantyl scaffold for biological applications stem from its inherent lipophilicity and its capacity to enhance drug stability, leading to improved pharmacokinetics of modified drug candidates. The rigid cage structure shields adjacent functional groups from metabolic cleavage, thereby augmenting drug stability and distribution in blood plasma [1].

Consequently, the preparation and testing of new derivatives of adamantane continue to be a highly pertinent focus in medicine, as well as in other fields. In this part, the structures of the adamantane-derived Schiff bases described so far, as well as their complex compounds, will be presented.

2. Adamantane-derived Schiff bases and their complexes

The Cambridge Structural Database search revealed a total of 224 structures of adamantane-derived Schiff bases, of which 131 complex compounds with *s*, *p*, and *d* metals are characterized by these azomethines. The most prevalent are complexes of Cu(II) (21 structures), Ti(IV) (19 structures), Zn(II) (18 structures), and Co(II) (11 structures). Additionally, six complexes with Pd(II) and Sn(II); five with Mn(II), four with Cu(I); three each with Fe(II), Ni(II), Zr(II), and Mg(II); two each with Ag(I), Au(I), Pt(II), In(III), V(IV), K(I), and Al(III); and one each with U(VI), Ca(II), Cr(III), Cr(II), Fe(III), Mn(0), Hf(II), Ir(I), Sb(II), Re(I), and Li(I) are known. In addition to these, mixed complexes are identified, including two complexes with Al(III) and Mo(0) (refcodes QIHVEM, QIHXEO), one with Cu(II) and Cu(I) (refcode EKANEL), and one with Ru(II) and Co(II) metal centers (refcode HEXBOD). Out of these 131 structures, adamantane-derived Schiff bases are coordinated in their bis-condensed form in 57 complexes. The most prevalent coordination mode is bidentate

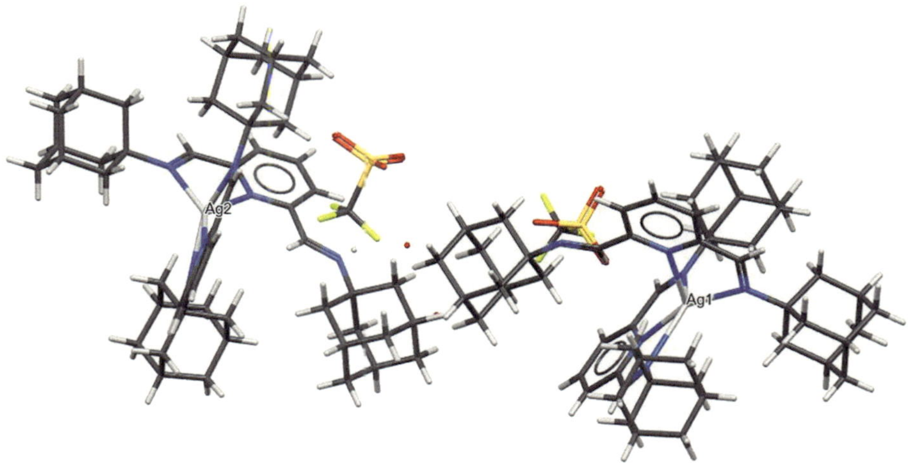

Figure 2.
The molecular structures of Ag(I) complex with 2,6-pyridyl-iminodiadamantane (refcode HAVJEW).

(80 structures), followed by monodentate and tridentate coordination in 22 complexes each, and tetradentate coordination in four complexes. In one complex structure, one part of the bis-condensed structure coordinates as bidentate, while the other coordinates as tridentate (refcode HAVJEW) [23]. In this Ag(I) complex compound (**Figure 2**), one 2,6-(pyridyl)- iminodiadamantane ligand adopts a tridentate coordination to the metal center, forming two planar five-membered chelate rings. The other ligand binds in a bidentate manner, resulting in a single five-membered chelate ring. In this case, the second imine N atom is anti-oriented to the pyridyl N atom, preventing chelate ring formation. The dihedral angles between the pyridyl ring and the adamantyl moieties connected to the imine N atoms in the tridentate chelate are similar. Conversely, the dihedral angle between the pyridyl ring and the adamantyl moiety linked to the imine N atom in the bidentate chelate is different. The adamantyl moiety connected to the imine N atom in this ligand remains non-coordinated [24].

The most common donor sets in examples of bidentate coordination are ON (42 complexes) and N_2 (28 complexes), followed by ON donor set from one part and NC donor set from the other part of the bis-condensed structure (three complexes), OS and NC donor sets (two complexes each), and NS, S_2, and O_2 donors in one complex each. Regarding tridentate coordination, donor sets ON_2 (six complexes), O_2N and N_3 (four complexes each), ONS (three complexes), ONP and N_2C (two complexes each), and ONC (one complex) are represented. Tetradentate donor sets in the examined structures include N_4 (two complexes), O_2N_2 (one complex), and ON_3 (one complex). It is noteworthy that dinuclear complexes (six structures), trinuclear and tetranuclear complexes (one structure each), and four structures of polynuclear complexes are structurally characterized [23].

It is interesting to note that among the investigated structures of Schiff bases with an adamantane ring, there are 8 structures with a pyrrole ring as well (refcodes EWOMEI, EWOMIM, LUHXIY, LUHZIA, PIJYAL, AVECUB, PEMDAQ, YARKUA), six with a piperidine/piperazine ring (refcodes HAWCUH, HAWDAO, IDOCIP, LANXEF, TANHAU, GOQJEE), three with a thiazole ring (refcodes PAWJEE, LANXEF, REJZIT), one with a furan ring (refcode FIBWEV), one with an imidazole ring (refcode BOQKOJ), one with a morpholine ring (refcode PIMHIF), and one with a nitrothiophene ring (refcode ZUQDAT) [23].

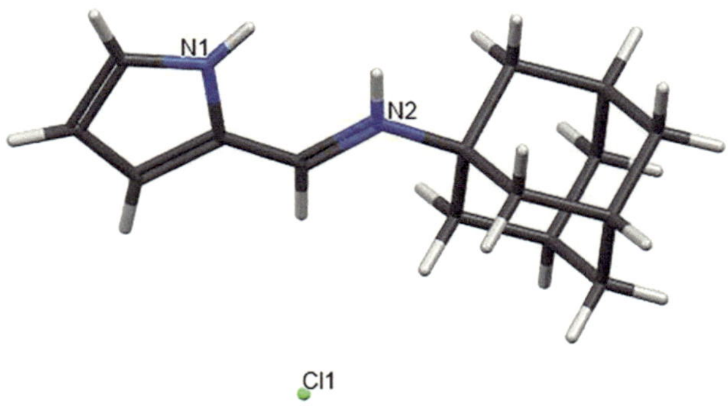

Figure 3.
Molecular structure of the 2-(1-adamantylimino)methyl-1H-pyrrole (refcode YARKUA).

The structure containing the pyrrole ring, 2-(1-adamantylimino)methyl-1H-pyrrole, which is presented in **Figure 3**, serves as a robust indicator for the convenient and highly accurate quantitative analysis of trace amounts of hydrochloric acid in various chlorinated organic solvents through UV-VIS absorption spectroscopy [25].

There are eight structures in which the adamantane-derived Schiff base ligand coordinates in the form of a zwitterion and one structure of a complex salt where the protonated form of Schiff base ([(E)-N-(adamantan-1-yl)(2H-1,3-benzodioxol-5-yl) methanimine]) appears as the cation, and the picrate ion serves as the anion (refcode WIRQEX, **Figure 4**) [23].

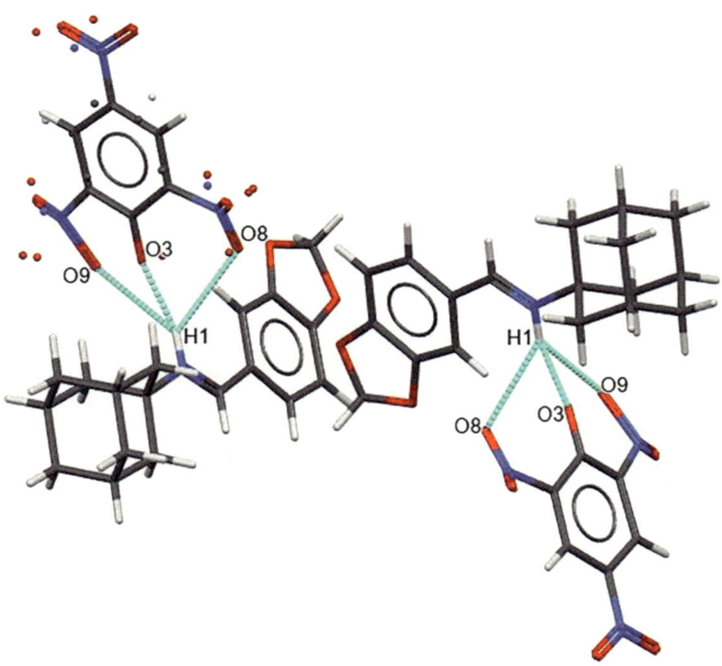

Figure 4.
The picrate salt containing a protonated Schiff base (E)-N-(adamantan-1-yl) (2H-1,3-benzodioxol-5-yl) methanimine with hydrogen bonds shown.

The ligand of interest, 2-[(E)-{3-[(E)-(adamantan-1-ylimino)methyl]-4-hydroxyphenyl}diazenyl]benzoic acid, in the work [26], has been utilized to synthesize two polymeric organotin complexes, with the ligand coordinating as a zwitterion. These adamantyl-substituted ligand derivatives demonstrate selective toxicity against HeLa cancer cells while being less toxic to healthy HEK 293 cells. These compounds induce apoptosis in HeLa cells, with the apoptotic effect increasing at higher concentrations, as confirmed by specialized biochemical assays. The presence of biocompatible adamantyl groups in compounds presented in **Figure 5** enhances the lipophilicity and stability of the molecular tin complexes, making them promising for future triorganotin antitumor therapeutics [26].

There is one adamantane-derived metallomacrocycle described in the work [27]. The work emphasizes the efficacy of bis(bidentate)-bis(N-acylamidines), incorporating various spacer units, in self-assembling supramolecular architectures. The ligand of interest features a sterically fixed, curved 1,3-adamantanediyl spacer with a tetrahedral spatial orientation, facilitating bis(coordination) in a parallel orientation. This ligand leads to the formation of a Ni(II) metallomacrocycle due to its specific curved structure. Two nickel atoms coordinate with ligands, creating square-planar *cis*-N,O chelates from deprotonated N-acylamidine moieties (**Figure 6**). Minor deviations from the ideal square-planar geometry are observed, with the Ni(II) ion

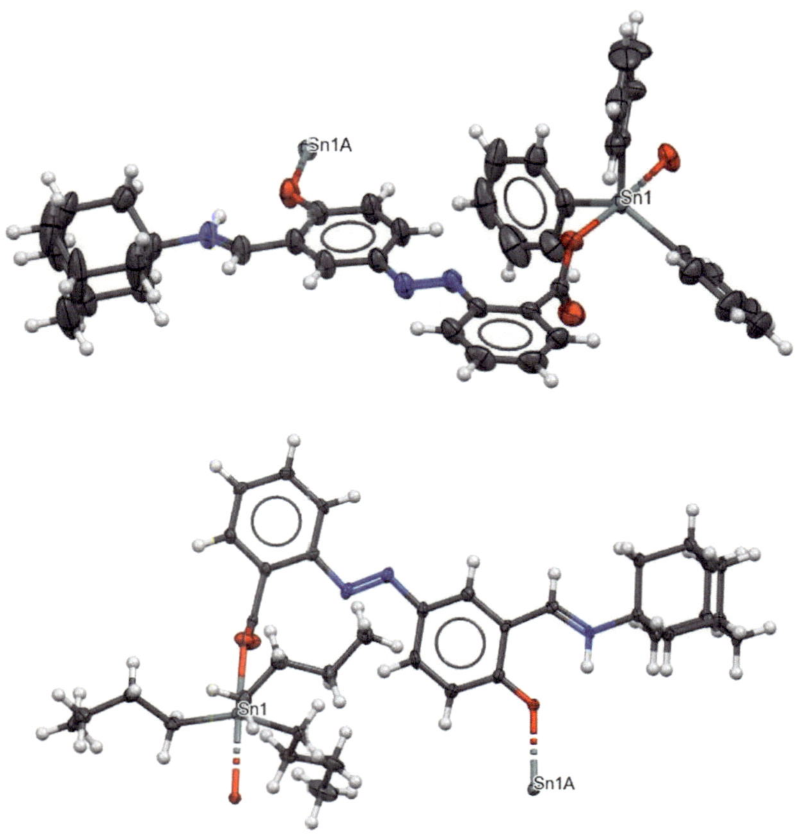

Figure 5.
The molecular structures of two polymeric organotin complexes where the ligand is 2-[(E)-{3-[(E)-(adamantan1-ylimino)methyl]-4-hydroxyphenyl}diazenyl]benzoic acid (refcode OMETAE – above, refcode OMETOS – below).

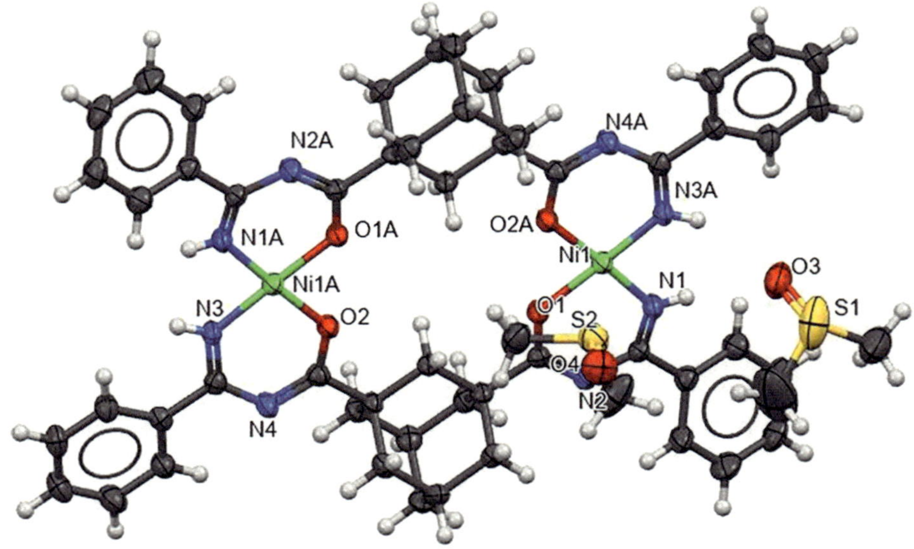

Figure 6.
The molecular structures of Ni(II) metallomacrocycle with adamantane-derived Schiff base (refcode TEYHEM).

positioned slightly above the plane. The size of the cavity is dictated by the distance between face-to-face adamantane units and the separation between two oxygen atoms of the same ligand [27].

3. Complexes of the adamantane-1-carbohydrazide derivatives

There are 30 structures of hydrazones and thiohydrazones Schiff bases containing an adamantane ring.

In the study of Leovac et al. the structural characterization of Cu(II) complexes with hydrazonic derivatives of adamantane was undertaken for the first time. Specifically, this paper describes the synthesis and characterization of two ligands and their corresponding five-coordinated square–pyramidal complexes [28].

The first ligand (Adpy) was obtained through the condensation reaction of 2-acetylpyridine and adamantane-1-carbohydrazide. Three complexes were characterized with this ligand, namely [$CuCl_2$(Adpy)], [$Cu_2(\mu\text{-}Cl)_2$(Adpy-H)$_2$], and [$Cu(NCS)_2$(Adpy)]. The structure for the first mentioned complex was validated through elemental analysis and infrared spectra, due to the unsuitability of the obtained crystals for X-ray, while the molecular structures of the remaining two complexes are presented in **Figure** 7. Compound [$Cu_2(\mu\text{-}Cl)_2$(Adpy-H)$_2$] is a centrosymmetrical dinuclear neutral complex with Cu(II) ions in deformed square-pyramidal environments. It features an ON_2-coordinated monoanionic organic ligand (Adpy-H) in the basal plane, and the remaining sites are occupied by coligands. In its crystal structure, two chlorido anions fill the fourth basal and fifth apical sites, forming a nearly rectangular $Cu_2(\mu\text{-}Cl)_2$ core. In the coordination sphere of monomeric neutral complex [$Cu(NCS)_2$(Adpy)], there are two thiocyanato ions with distinct coordination modes. One is situated in the basal plane, coordinated through a nitrogen atom, while the other occupies the apical position of the square pyramid, coordinated through a sulfur atom. This work states that this compound stands out as an

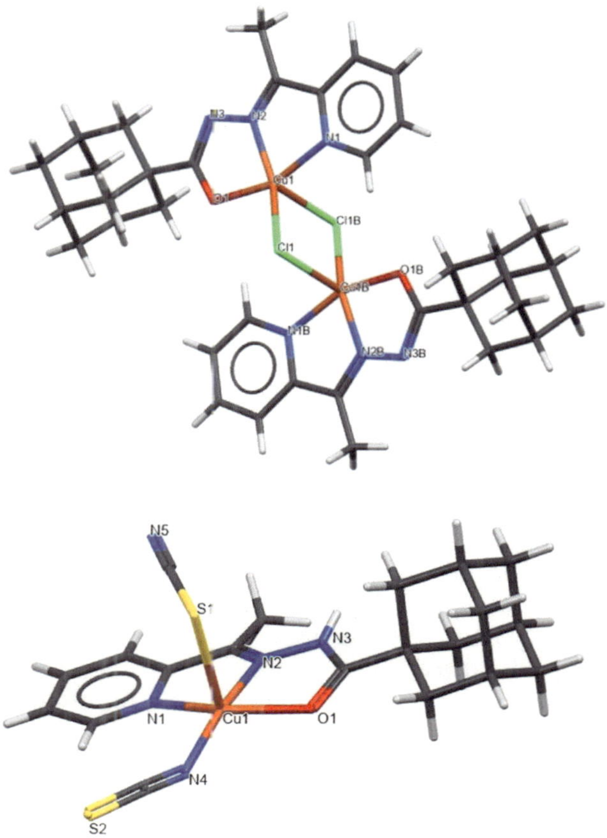

Figure 7.
The molecular structures of the [Cu2(μ-Cl)2(Adpy-H)2] (refcode WOTDAM) – above, and [Cu(NCS)2(Adpy)] (refcode WOTDEQ) – below.

instance of a rare category of mononuclear complexes characterized by the formula [CuL(NCS)(SCN)], where L denotes a tridentate ligand [28].

The second ligand (Addpy) was obtained through the condensation reaction of di(2-pyridyl) ketone and adamantane-1-carbohydrazide. With this ligand, an additional two centrosymmetrical dinuclear neutral complexes were characterized, namely $[Cu_2(\mu\text{-}Cl)_2(Addpy\text{-}H)_2]$ and $[Cu_2(NCS)_2(\mu\text{-}Addpy\text{-}H)_2]$ (**Figure 8**). For $[Cu_2(\mu\text{-}Cl)_2(Addpy\text{-}H)_2]$, the fourth basal and fifth apical positions in square – pyramide polyhedron are occupied by two chlorido anions that are crystallographically related through an inversion operation. Consequently, this complex also displays a nearly rectangular $Cu_2(\mu\text{-}Cl)_2$ core, resembling the structural pattern found in other copper(II) complexes utilizing ON_2-tridentate hydrazones. For $[Cu_2(NCS)_2(\mu\text{-}Addpy\text{-}H)_2]$, the fourth basal position is filled by an N-coordinated thiocyanato ion, while the apical site is occupied by the N4 pyridinic nitrogen atom of a tetradentate μ-Addpy ligand. In this work, it is stated that this behavior mirrors what has been reported for di(2-pyridyl) ketone Schiff bases [28].

The anticancer potential, as well as the DNA binding and cleavage activity, of these five complexes has been assessed. These complexes displayed strong cytotoxicity, triggering apoptosis as the primary method of cell death. DNA-binding activities were investigated through UV/Vis absorption spectroscopy and fluorescence emission

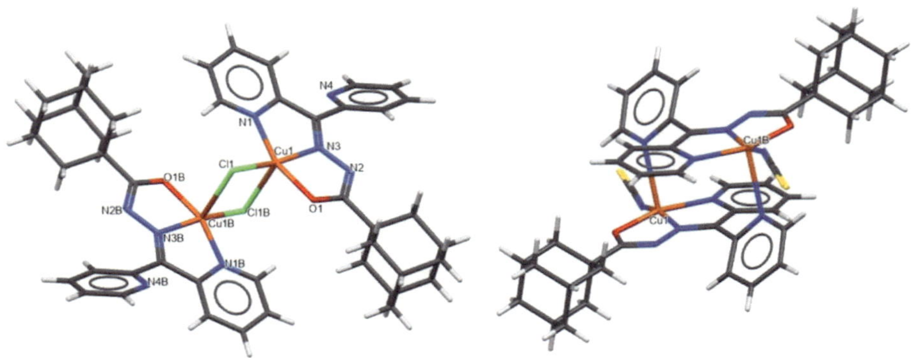

Figure 8.
The molecular structures of the $[Cu_2(\mu\text{-}Cl)_2(Addpy\text{-}H)_2]$ *(refcode WOTDIU) – Left, and* $[Cu_2(NCS)_2(\mu\text{-}Addpy\text{-}H)_2]$ *(refcode WOTDOA) – Right.*

measurements, indicating an intercalative mode of interaction. Additionally, the complexes caused significant double-strand cleavage of supercoiled DNA [28].

With the mentioned ligand Addpy, Rodić and collaborators synthesized and characterized three additional complexes in their study, with the formulas $[Cu^{II}_2Cu^{I}_2(Addpy)_2Br_2(\mu\text{-}Br_4)]$, *catena*-poly $[CuCl(\mu Addpy)(\mu\text{-}Cl)CuCl_2]_n$, and $[Cu(Addpy)(NCS)_2]$ (**Figure 9**) [29].

In the centrosymmetrical tetranuclear complex $[Cu^{II}_2Cu^{I}_2(Addpy)_2Br_2(\mu\text{-}Br_4)]$, copper atoms have distinct coordination environments and oxidation states. Copper atom with oxidation state +2 (Cu1) is placed in a deformed square-pyramidal environment. In this polyhedron, ON_2 donor atoms of the neutral ligand and bromide ion are in the basal plane, and a second bromide ion is in the apical position (this bromide is a linkage between Cu1 and Cu_2). Copper atom with the oxidation state +1 (Cu_2) is placed in a trigonal-planar geometry by three bridging bromide ions [29].

In the polymeric complex from Rodić and collaborators' work, the Addpy coordinates to Cu1 in its neutral form as tetradentate ON_3-ligand, using two nitrogen atoms and one oxygen atom to bind with Cu1 and one nitrogen atom from the second pyridine ring to bind with Cu2. The chloride anion and the ON_3-coordinated Addpy play bridging roles between two copper atoms [29].

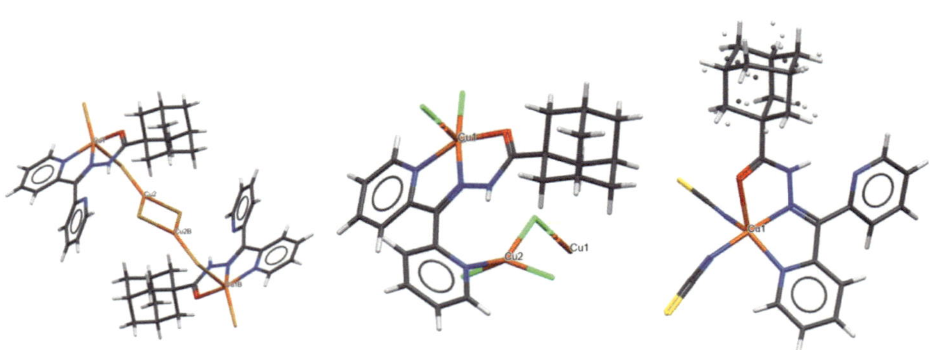

Figure 9.
The molecular structures of the $[Cu^{II}_2Cu^{I}_2(Addpy)_2Br_2(\mu\text{-}Br_4)]$ *(refcode EKANEL) – The first, catena-poly* $[CuCl(\mu Addpy)(\mu\text{-}Cl)CuCl_2]_n$ *(refcode EKANIP) – The second, and* $[Cu(Addpy)(NCS)_2]$ *(refcode EKANOV) – The third presented structure (from left to right).*

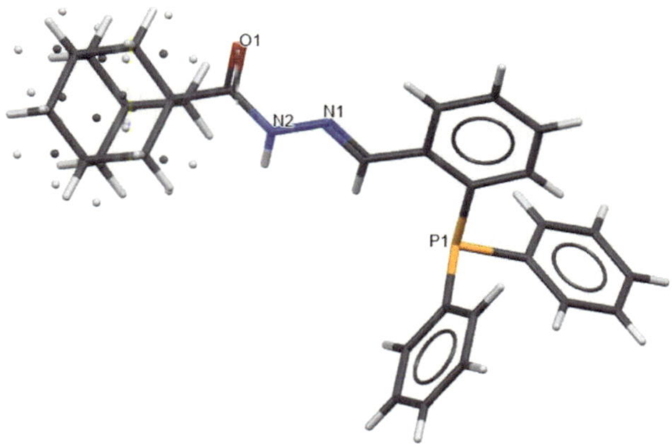

Figure 10.
Molecular structure of the 2-(diphenylphosphino) benzaldehyde-1-adamantoylhydrazone (refcode SIVXUS).

In the mononuclear complex [Cu(Addpy)$(NCS)_2$], copper(II) atom is placed in a deformed square-pyramidal environment of ON_2 donor atoms of the Addpy ligand and one thiocyanate ion in the basal plane and another thiocyanate ion in apical position. For this coordination polyhedron, a noteworthy deviation of the Cu(II) ion from the basal donor plane is observed – 0.3894(5) Å [29].

Outperforming cisplatin in terms of *in vitro* cytotoxicity against some cancer cell lines, these Cu(II) complexes demonstrate strong anticancer activity. These compounds trigger apoptosis in HeLa cells and exhibit angiogenesis inhibition in vascular endothelial cells [29].

The study of Đorđević et al. describes the synthesis and crystal structure of the 2-(diphenylphosphino)benzaldehyde 1-adamantoylhydrazone (HL) (**Figure 10**) and its two complexes [30].

This ligand reacts with $K_2[MCl_4]$ (M = Pd(II), Pt(II)) in ethanol, at 50°C for 1 h, yielding square-planar neutral complexes [M(L)Cl] (**Figure 11**). These complexes

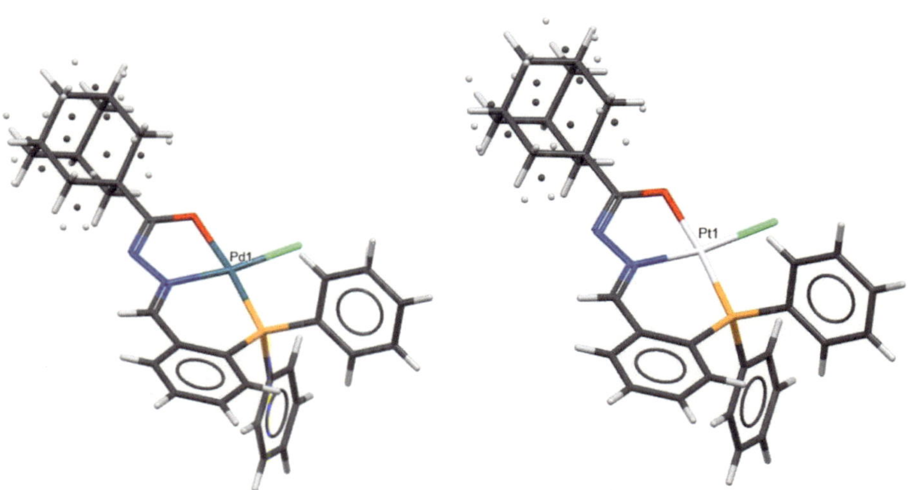

Figure 11.
The molecular structures of the [Pd(L)Cl] (refcode SIVXOM) – Left, and [Pt(L)Cl] (refcode SIVXIG) – Right (L = 2-(diphenylphosphino)benzaldehyde-1-adamantoylhydrazone).

feature tridentate ONP coordination from the monoprotonated Schiff base and a chloride ligand. Their antimicrobial and cytotoxic properties were assessed against human larynx carcinoma cells (Hep-2) and non-cancerous human lung fibroblast cells (MRC-5). The platinum(II) complex exhibits cytotoxicity toward Hep-2 cells comparable to oxaliplatin, with enhanced selectivity for cancer cells [30].

In the reaction of 2,6-diacetylpyridine and adamantane-1-carbohydrazide in ethanol, in malor ratio 1:2, under the reflux, a Schiff base was obtained – $(Ad)_2dap$ (**Figure 12**) [31].

A dinuclear complex of Zn(II), $[Zn_2((Ad)_2dap{-}H)_2]{\bullet}EtOH$ (**Figure 13**), was synthesized by reacting a warm suspension of this ligand in ethanol with a warm ethanolic solution of $Zn(CH_3COO)_2$ in a 1:1 molar ratio. The Schiff base in the synthesized complex compound acts as a pentadentate ligand with O_2N_3 coordination.

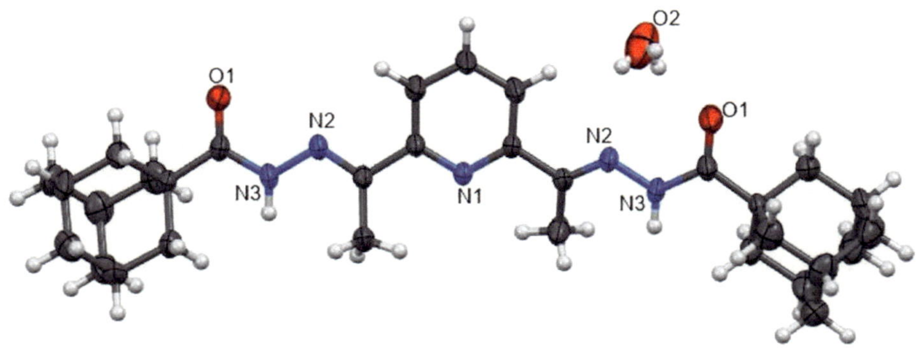

Figure 12.
The molecular structure of the $(ad)_2dap$.

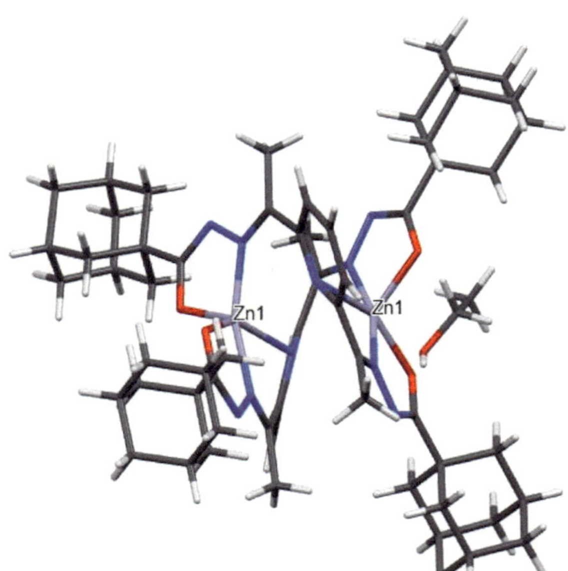

Figure 13.
The molecular structure of the $[Zn_2((ad)_2dap{-}H)_2]{\cdot}EtOH$ ($(ad)_2dap$ = 2,6-diacetylpyridine-adamantane-1-carbohydrazide).

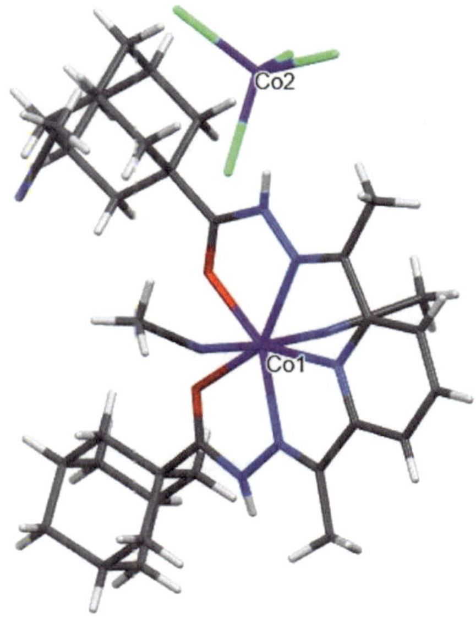

Figure 14.
The molecular structure of the[Co(ad)$_2$dap(CH3CN)$_2$][CoCl$_4$]·CH$_3$CN ((ad)$_2$dap = 2,6-diacetylpyridine-adamantane-1-carbohydrazide).

Consequently, six five-membered metallocycles were formed. Zinc occupied a slightly deformed square-pyramidal environment (τ_5 = 0.24), with donor nitrogen atoms and one oxygen atom situated in the equatorial plane and another oxygen atom positioned in the apical position [31].

In the reaction of a warm acetronitrile suspension of this ligand and a warm acetone solution of cobalt(II) chloride, in a molar ratio 1:2, green crystals of a complex compound of the formula [Co(Ad)$_2$(CH$_3$CN)$_2$dap][CoCl$_4$]•CH$_3$CN (**Figure 14**) were obtained. The ligand, in its neutral form, is coordinated to the central Co(II) as O_2N_3 pentadentate, resulting in the formation of four five-membered metallocycles. Cobalt(II) is placed in pentagonal bipyramidal environment with the ligand's donor atoms at the base of this polyhedron and acetonitrile nitrogen in apical positions [32].

4. Complexes of the adamantyl semicarbazone and benzoyl derivatives

In the study [33], D. Palanimuthu and collaborators detail the synthesis and characterization of two groups of adamantane-based compounds – adamantyl semicarbazones and adamantyl benzoyl hydrazones, as well as some Cu(II) and Fe(III) complexes formed with them (12 structures).

This research focuses on developing novel adamantane-based semicarbazones and hydrazones targeting multiple pathological hallmarks associated with Alzheimer's disease (AD). Compound derived from 2-pyridinecarboxaldehyde (N-adamantan-1-yl)benzoyl-4-amidohydrazone emerged as a lead compound, demonstrating efficacy in iron chelation, attenuation of CuII-mediated β-amyloid aggregation, low cytotoxicity, inhibition of oxidative stress, and favorable blood–brain barrier permeation. The study suggests these multi-functional agents present a promising therapeutic strategy for AD treatment [33].

The molecular structure of the ferric complex from this work is illustrated in **Figure 15**. The adamantane-derived ligand, with both parts deprotonated at the phenol O-atom, coordinates in a tridentate meridional manner, forming a monocationic distorted octahedral complex. This represents the first crystallographically characterized phenolic semicarbazone complex of Fe(III). The Fe—O bond lengths and Fe—N bond lengths align with a high-spin d^5 electronic configuration, akin to ferric complexes of related O_2N phenolic hydrazone ligands [33].

The copper complex published within the same research displays a distorted square pyramidal geometry (**Figure 16**). The asymmetric unit contains two complexes and 1.5 MeOH molecules. This complex features a basal plane occupied by the tridentate ligand's O_2N donor atoms, along with a chlorido ligand, while weakly coordinated chlorido ligands occupy the axial sites. The ligand is coordinated as a zwitterion, with protonated pyridoxal N atom and deprotonated pyridoxal OH group. The presence of partially occupied MeOH correlates with the disorder of the hydroxymethyl substituent on the pyridyl ring. The axially elongated square pyramidal (pseudo-square planar) coordination geometry found in this structure is also observed in other mononuclear copper complexes of tridentate semicarbazone ligands with phenolic O donors [33].

The other copper(II) complex (**Figure 17**) crystallizes as a centrosymmetric dimer with two copper ions bridged by two phenolate O-donors. Each copper(II) ion adopts a square pyramidal geometry, with the tridentate O_2N semicarbazone ligand and the phenolate O donor atom of the adjacent complex occupying the equatorial plane, and the axial site is occupied by a chlorido ligand. Intermolecular hydrogen bonding was also observed, with the chlorido ion strongly hydrogen bonding to two NH groups of an adjacent molecule. This bis-phenolate bridged di-Cu^{II} coordination mode has been noted in various analogous complexes of semicarbazone ligands [33].

This study introduces the first instances of adamantane-based semicarbazone and hydrazone chelators showcasing multifunctional activity for Alzheimer's disease

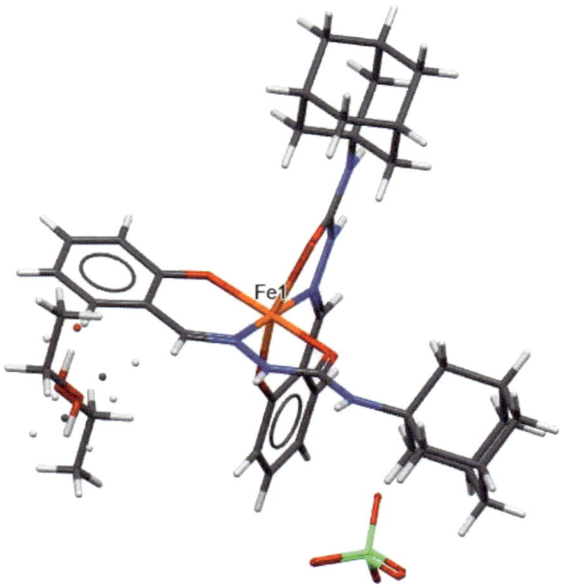

Figure 15.
The molecular structure of the Fe(III) complex with salicylaldehyde-N-adamantan-1- ylsemicarbazide (refcode LEYGII).

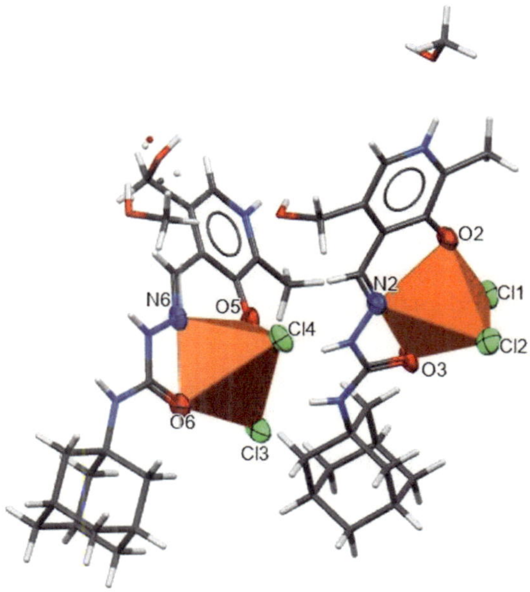

Figure 16.
The molecular structure of the Cu(II) complex with pyridoxal-N-adamantan-1-yl-semicarbazone (refcode LEYGAA).

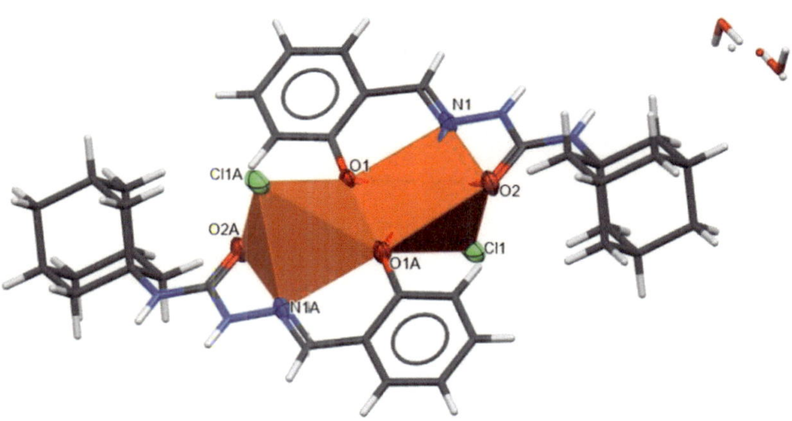

Figure 17.
The molecular structure of the Cu(II) complex with pyridoxal-N-adamantan-1-yl-semicarbazone (refcode LEYGEE).

(AD) treatment. The adamantoyl semicarbazone and adamantoyl benzoyl hydrazone analogues, along with the pyridil derivative outlined in this research, collectively exhibit advantageous multi-functional properties aimed at addressing the various hallmarks of AD [33].

5. Conclusion

In conclusion, the investigation of adamantane-derived Schiff bases and their complexes, conducted through the Cambridge Structural Database search, sheds light on a promising avenue for drug discovery and therapeutic development. The unique chemical properties of adamantane, including its hydrophobic nature and metabolic stability, make it an attractive scaffold for modifying drug molecules across various therapeutic classes. The demonstrated ability of these derivatives to disrupt enzymes and exhibit diverse therapeutic activities underscores the importance of synthesizing and characterizing novel adamantane-based compounds. This overview not only contributes to the advancement of medicinal chemistry but also provides valuable insights into chemical synthesis and materials science. These insights highlight the potential for future innovations in drug design and development, informed by a robust understanding of adamantane chemistry and its therapeutic implications.

Acknowledgements

The authors gratefully acknowledge the financial support of the Ministry of Science, Technological Development and Innovation of the Republic of Serbia (Grants No. 451-03-66/2024-2103/200125 & 451-03-65/2024-2103/200125).

Conflict of interest

The authors declare no conflict of interest.

Author details

Marijana S. Kostić* and Vukadin M. Leovac
Faculty of Sciences, University of Novi Sad, Novi Sad, Serbia

*Address all correspondence to: marijana.kostic@dh.uns.ac.rs

References

[1] Štimac A, Šekutor M, Mlinarić-Majerski K, Frkanec L, Frkanec R. Adamantane in drug delivery systems and surface recognition. Molecules. 2017;**22**(2):297-311. DOI: 10.3390/molecules22020297

[2] Calıs Ü, Yarım M, Köksal M, Özalp M. Synthesis and antimicrobial activity evaluation of some new adamantane derivatives. Arzneimittel-Forschung/Drug Research. 2002;**52**(10):778-781. DOI: 10.1002/chin.200307067

[3] Wanka L, Iqbal K, Schreiner PR. The lipophilic bullet hits the targets: Medicinal chemistry of adamantane derivatives. Chemical Reviews. 2013;**113**(5):3516-3604. DOI: 10.1021/cr100264t

[4] Gerzon K, Krumkalns EV, Brindle RL, Marshall FJ, Root MA. The adamantyl group in medicinal agents. I. Hypoglycemic N-arylsulfonyl-N'-adamantylureas. Journal of Medicinal Chemistry. 1963;**6**(6):760-763. DOI: 10.1021/jm00342a029

[5] Rapala RT, Kraay RJ, Gerzon K. The adamantyl group in medicinal agents. II. Anabolic steroid 17β-adamantoates. Journal of Medicinal Chemistry. 1965;**8**(5):580-583. DOI: 10.1021/jm00329a007

[6] Gerzon K, Tobias DJ, Holmes RE, Rathbun RE, Kattau RW. The adamantyl group in medicinal agents. IV. Sedative action of 3,5,7-trimethyladamantane-1-carboxamide and related agents. Journal of Medicinal Chemistry. 1967;**10**(4):603-606. DOI: 10.1021/jm00316a018

[7] Van Dijk J, Zwagemakers JM. Oxime ether derivatives, a new class of nonsteroidal antiinflammatory compounds. Journal of Medicinal Chemistry. 1977;**20**(9):1199-1206. DOI: 10.1021/jm00219a018

[8] Burger A, Wolff ME. Burger's Medicinal Chemistry, Part II. New York: Wiley; 1981. 557 p

[9] Bailey EV, Stone TW. The mechanism of action of amantadine in parkinsonism: A review. Archives internationales de pharmacodynamie et de therapie. 1975;**216**:246-262

[10] Tsitsa P, Antoniadou-Vyza E, Hamodrakas SJ, Eliopoulos EE, Tsantili-Kakoulidou A, Roussakis C. Synthesis, crystal structure and biological properties of a new series of lipophilic s-triazines, dihydrofolate reductase inhibitors. European Journal of Medicinal Chemistry. 1993;**28**:149-158. DOI: 10.1016/0223-5234(93)90007-2

[11] Antoniadou-Vyza E, Tsitsa P, Hytiroglou E, Tsantili-Kakoulidou A. New adamantan-2-ol and adamantan-1-methanol derivatives as potent antibacterials. Synthesis, antibacterial activity and lipophilicity studies. European Journal of Medicinal Chemistry. 1996;**31**:105-110. DOI: 10.1016/0223-5234(96)80443-0

[12] Cody V. In: Rein R, editor. Molecular Basis of Cancer, Part II. New York: Alan R. Liss; 1984. 275 p

[13] Ajaz A, Shaheen AM, Ahmed M, Munawar SK, Siddique AB, Karim A, et al. Synthesis of an amantadine-based novel Schiff base and its transition metal complexes as potential ALP, a-amylase, and a-glucosidase inhibitors. RSC Advances. 2023;**13**:2756-2767. DOI: 10.1039/d2ra07051k

[14] Vamecq J, Van Derpoorten K, Poupaert JH, Balzarini J, De Clercq E, Stables JP. Anticonvulsant phenytoi nergic pharmacophores and anti-HIV activity- preliminary evidence for the dual requirement of the 4-aminophthalimide platform and the N-(1-adamantyl) substitution for antiviral properties. Life Sciences. 1998;**63**(19):267-274. DOI: 10.1016/s0753-3322(97)82327-x

[15] Ilies MA, Masereel B, Rolin S, Scozzafava A, Câmpeanu G, Cîmpeanu V, et al. Carbonic anhydrase inhibitors: Aromatic and heterocyclic sulfonamides incorpo rating adamantyl moieties with strong anticonvulsant activity. Bioorganic & Medicinal Chemistry. 2004;**12**(10):2717-2726. DOI: 10.1016/j.bmc.2004.03.008

[16] Lamoureux G, Artavia G. Use of the adamantane structure in medicinal chemistry. Current Medicinal Chemistry. 2010;**17**(26):2967-2978. DOI: 10.2174/092986710792065027

[17] Zoidis G, Papanastasiou I, Dotsikas I, Sandoval A, Dos Santos RG, Papadopoulou-Daifoti Z, et al. The novel GABA adamantane derivative (AdGABA): Design, synthesis, and activity relationship with gabapentin. Bioorganic & Medicinal Chemistry. 2005;**13**(8):2791-2798. DOI: 10.1016/j.bmc.2005.02.030

[18] Antonov SM, Johnson JW, LukomskayaNY, PotapyevaNN, GmiroVE, Magazanik LG. Novel adamantane derivatives act as blockers of open ligand-gated channels and as anticonvulsants. Molecular Pharmacology. 1995;**47**(3):558-567

[19] Tanner JA, Zheng BJ, Zhou J, Watt RM, Jiang JQ, Wong KL, et al. The adamantane-derived bananins are potent inhibitors of the helicase activities and replication of SARS coronavirus. Chemical Biology. 2005;**12**(3):303-311. DOI: 10.1016/j.chembiol.2005.01.006

[20] Cimolai N. Potentially repurposing adamantanes for COVID-19. Journal of Medical Virology. 2020;**92**(6):531-532. DOI: 10.1002/jmv.25752

[21] Mathur A, Beare AS, Reed SE. *In vitro* antiviral activity and pre liminary clinical trials of a new adamantane compound. Antimicrobial Agents and Chemotherapy. 1973;**4**(4):421-426. DOI: 10.1128/AAC.4.4.421

[22] Wu Z, Wei W, Ma J, Luo J, Zhou Y, Zhou Z, et al. Adsorption of iodine on adamantane-based covalent organic frameworks. Chemistry Select. 2021;**6**:10141-10148. DOI: 10.1002/slct.202102656

[23] Groom CR, Bruno IJ, Lightfoot MP, Ward SC. The Cambridge structural database. Acta Crystallographica. 2016;**B72**:171-179. DOI: 10.1107/S2052520616003954

[24] Jimenez J, Chakraborty I, Del Cid AM, Maschara PK. Five- and six-coordinated silver(I) complexes derived from 2,6-(Pyridyl) iminodiadamantanes: Sustained release of bioactive silver toward bacterial eradication. Inorganic Chemistry. 2017;**56**(9):4784-4787. DOI: 10.1021/acs.inorgchem.7b00621

[25] Anzai K, Kawamorita S, Komiya N, Naota T. Convenient spectroscopic method for quantitative analysis of trace hydrochloric acid in chlorinated organic solvents using 2-(1-adamantylimino) methyl-1H-pyrrole as a robust indicator. Chemistry Letters. 2017;**46**(5):672-675. DOI: 10.1246/cl.170065

[26] Basu Baul TS, Rao Addepalli M, Duthie A, Singh P, Koch B, Gildenast H,

et al. Triorganotin(IV) derivatives with semirigid heteroditopic hydroxo-carboxylato ligands: Synthesis, characterization, and cytotoxic properties. Applied Organometallic Chemistry. 2020;**35**(2):e6080-e6107. DOI: 10.1002/aoc.6080

[27] Clodt JI, Fröhlich R, Eul M, Würthwein EU. Metallomacrocyclic complexes by self-assembly of NiII and CuII ions and chelating bis(N-acylamidines). European Journal of Inorganic Chemistry. 2012;**2012**(8): 1210-1217. DOI: 10.1002/ejic.201101350

[28] Leovac VM, Rodić MV, Jovanović LS, Joksović MD, Stanojković T, Vujčić M, et al. Transition metal complexes with 1-adamantoyl hydrazones – cytotoxic copper(II) complexes of tri- and tetradentate pyridine chelators containing an adamantane ring system. Journal of Inorganic Chemistry. 2015;**2015**(5):882-895. DOI: 10.1002/ejic.201403050

[29] Rodić MV, Leovac VM, Jovanović LS, Spasojević V, JoksovićMD, StanojkovićT, et al. Synthesis, characterization, cytotoxicity and antiangiogenic activity of copper(II) complexes with 1-adamantoyl hydrazone bearing pyridine rings. European Journal of Medicinal Chemistry. 2016;**115**:75-81. DOI: 10.3390/molecules25112530

[30] Đorđević M, Jeremić D, Rodić MV, Simić V, Brčeski I, Leovac VM. Synthesis, structure and biological activities of Pd(II) and Pt(II) complexes with 2-(diphenylphosphino)benzaldehyde 1-adamantoylhydrazone. Polyhedron. 2014;**68**:234-240. DOI: 10.1016/j.poly.2013.10.029

[31] Kostić MS, Leovac VM, Bogdanović MG, Rodić MV, Vojinović Ješić LS, Radanović MM. Synthesis and characterization of the Schiff base 2,6-diacetylpyridine-adamantane-1-carbohydrazide and its zinc(II) complex. In: Abstracts of the XXVIII Conference of the Serbian Crystallographic Society; 14-15th June 2023, Čačak, Serbia. Belgrade: Serbian Crystallographic Society; 2023. pp. 16-17

[32] Kostić MS, Bogdanović MG, Radanović MM. Synthesis and characterization of the Co(II) complex with 2,6-diacetylpyridine-adamantane-1-carbohydrazide Schiff base. In: Book of Abstracts of the 9th Conference of Young Chemists of Serbia; 4th November 2023, Novi Sad, Serbia. Belgrade: Serbian Chemical Society; 2023. p. 65

[33] Palanimuthu D, Wu Z, Jansson PJ, Braidy N, Bernhardt PV, Richardson DR, et al. Novel chelators based on adamantane-derived semicarbazones and hydrazones that target multiple hallmarks of Alzheimer's disease. Dalton Transactions. 2018;**47**:7190-7205. DOI: 10.1039/C8DT01099D

Chapter 2

Chelating Agents in the Oilfield

Tariq Almubarak and Clarence Ng

Abstract

The focus in this chapter will be on the chemical subset of chelating agents commonly used in the petroleum industry. It will start by defining the functionality known as chelation. It will then share the possible applications of these chelating agents. Such applications include utilizing them as straight acidizing fluids, iron control agents, and inorganic scale removers. The chapter will then compare the corrosivity of these molecules towards typical metals used in industry. Finally, it will touch on the environmental aspect by sharing insights on the thermal degradation profile of the used chelating agents and its impact on nature.

Keywords: aminopolycarboxylic acid, chelating agent, acidizing, scale removal, iron control

1. Introduction

Chelating agents are compounds having at least two groups, usually referred to as ligators, that can transfer electrons to generate coordination bonds with a central metal atom. Due to the broad definition of the term, many compounds containing two or more electron donating groups, such as oxalates, fall into this category. However, in the oil and gas industry, the term "chelating agents" is often used to refer to a particular subclass of chelating agents known as aminopolycarboxylic acids (APCA). As the name suggests, these chelating agents contain one or more amine groups which form multiple metallocyclic rings from a single molecule via the formation of numerous coordinate bonds (**Figure 1**). In this review, we aim to provide readers with a basic understanding of chelation chemistry and the many applications of chelating agents in the oil and gas industry.

1.1 Stability constants

The equilibrium constant is an indication to whether the products of a reaction will remain stable in the solution or revert to its reactants. In the case of APCAs, this is typically referred to as the *stability constant* as it serves as an indication of the stability of the chelated product. The stability of a chelated product improves with a higher stability constant. Eqs. (1) and (2) can be used to determine this constant.

$$M + Y \rightleftarrows MY. \qquad (1)$$

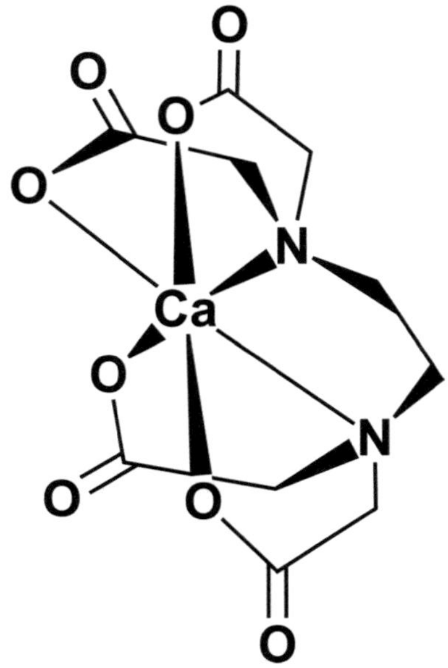

Figure 1.
Chelating agent rings in a Ca-EDTA complex [1].

$$\textit{Stability Constant} = \log\left(\frac{[MY]}{[\mathrm{M}][\mathrm{Y}]}\right). \quad (2)$$

Eq. (2) displays the stability constant in relation to the reactant and product concentrations in mol/L where M and Y react to produce the product MY. Additionally, this constant has been shown to rely on numerous factors, including the size of the ring created during chelation, the number of rings established, the basicity of the chelating agent, the type of donor atoms, and the central metal atom [2, 3]. For most chelating agents, a chelate ring size of five atoms offers the most stable structure, and any other chelation ring size would result in a loss in the structure's stability [4]. Since it is impossible to generalize the extent to which each of these specific characteristics affects the stability constant for all chelating agents, a combination of these factors can result in variations in the stability constants. Understanding the stability constants is an important first step to make when deciding on the appropriate chelating agent to be used in the field; selecting a chelating agent with low affinity towards the type of metal ion to be complexed would result in a failed treatment.

Examples of the chelating agents used in the oil industry include: nitrilo triacetic acid (NTA), hydroxyethylimino diacetic acid (HEIDA), ethylenediamine tetraacetic acid (EDTA), hydroxyethylethylenediamine triacetic acid (HEDTA), diethylenetriamine pentaacetic acid (DTPA), L-glutamic acid N, N-diacetic acid (GLDA), methylglycine diacetic acid (MGDA), 1,4,7,10-tetraazacyclododecane-1,4,7,10-tetraacetic acid (DOTA), and *trans*-1,2-cyclohexylenediamine tetraacetic acid (CDTA). A table of stability constants for these chelating agents can be found in **Table 1.**

Of these chelating agents, the most used are NTA, HEIDA, EDTA, HEDTA, DTPA, and GLDA [9, 10]. As can be seen in **Figure 2**, they are made up of between

	Ca^{2+}	Mg^{2+}	Fe^{2+}	Fe^{3+}	Ba^{2+}	Sr^{2+}	Al^{3+}	Zn^{2+}	Zr^{4+}
NTA	6.4	5.4	8.8	15.8	4.8	5.0	11.4	10.6	20.8
HEIDA	4.8	3.5	6.8	11.6	2.8	3.8	7.7	8.4	
EDTA	10.7	8.7	14.3	25.7	7.7	8.6	16.1	17.5	29.9
HEDTA	8.4	7.0	12.2	19.8	6.2	6.9	15.6	14.7	
DTPA	10.9	9.3	16.5	28.0	8.6	9.7	18.4	18.8	36.9
GLDA	5.9	5.2	8.7	15.2	3.5	4.1	12.2	11.5	
MGDA	7.0	5.8	8.1	16.5	4.9	5.2		10.9	
DOTA	17.2	11.8	20.2	29.4	12.9	15.0	17.0	20.5	
CDTA	13.1	11.1	18.9	30.0	8.6	10.5	19.5	19.3	29.9

Table 1.
Stability constants at 77°F (25°C) [5–8].

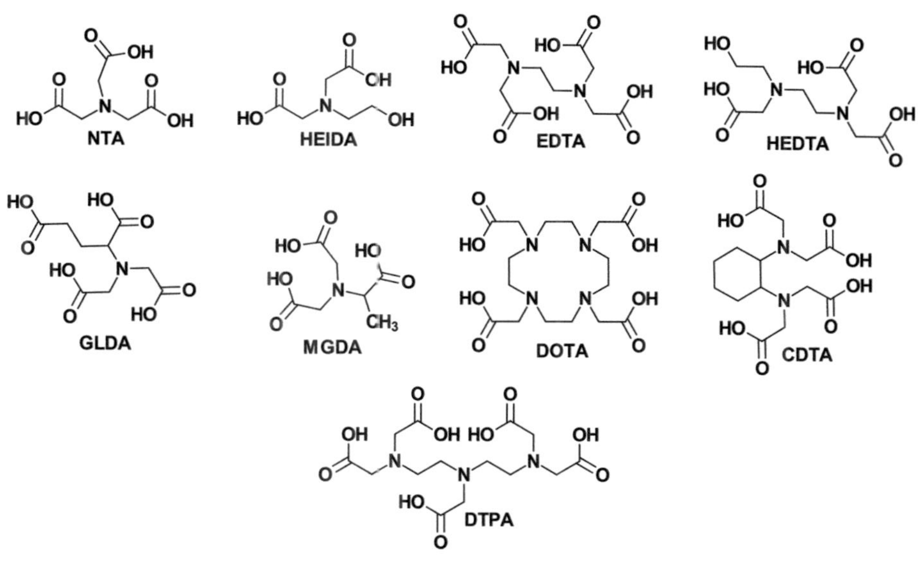

Figure 2.
Structural formulas of commonly used aminopclycarboxylic acids in the petroleum industry [11, 12].

one and three amine groups and multiple carboxylic arms that branch off the amine. Readers may notice that several of these chelating agents have similar structural formulas, such as EDTA and HEDTA. This is due to the addition of a hydroxyl functional group in place of a carboxylate functional group to improve the water solubility of the chelating agent. While this replacement comes at the cost of its stability constant (**Table 1**), HEDTA does not face any solubility problems while EDTA tends to precipitate at low pH. These APCAs can be described by the general chemical formula HxY, where x depends on how many carboxylate groups are protonated in the molecule. At higher pH levels, the APCA can be deprotonated until all protons are removed. The coordination interactions between the amine/carboxylate groups of deprotonated APCA and the metal ion then serve to chelate the metal ion [13].

1.2 Mechanism of mineral dissolution

One important application of chelating agents in the oilfield is to dissolve formation rock and scale deposits. Therefore, it is necessary to understand the method of mineral dissolution. The rock and scale deposits are found in the wellbore or in the tubulars. These minerals are often covered with oil leaving them oil wet. This necessitates several preparation stages to remove the oily layers and to generate a water wet surface that would allow for the dissolution mechanisms discussed in this section.

The method of chelating agent attack on rocks such as calcite or dolomite, happens through a mechanism called chelating agent assisted dissolution. At a high pH, solution coordination would be the primary dissolution mechanism while surface complexation would be prevalent at lower pH [14–16]. The pH of the chelating agent can be adjusted by the addition of strong acids and bases such as HCl and NaOH. The optimal pH is in the range of 3–5 for low pH applications and above 10 for high pH applications.

While solution coordination involves the physical dissolution of the metal ion before it is chelated in solution (**Figure 3**), surface complexation involves a different process as described by Nowack and Sigg [17] and Fredd and Fogler [18].

The surface complexation mechanism can be described in five steps as seen in (**Figure 4**).

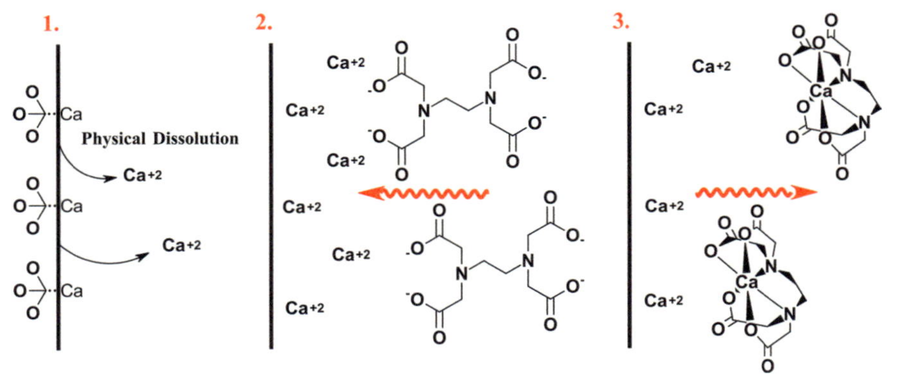

Figure 3.
Representation of solution coordination mechanism of EDTA chelating agent (more common at high pH) [1].

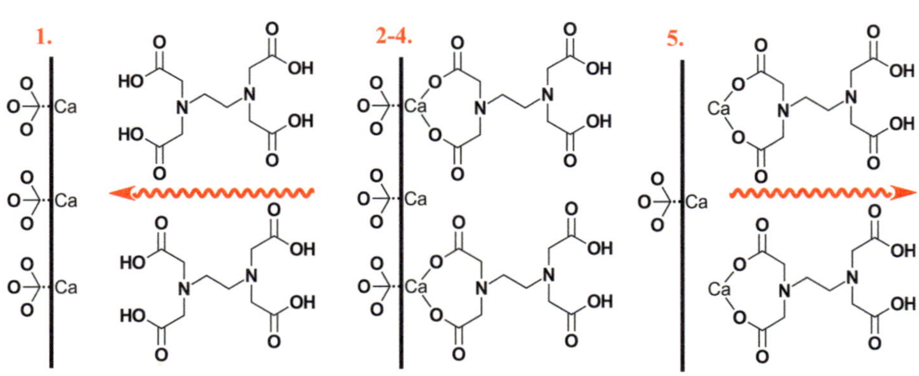

Figure 4.
Representation of surface complexation mechanism of EDTA chelating agent (more common at low pH) [1].

1. Chelating agent diffusion from solution to the surface of the mineral.

2. The initial adsorbed compound is formed.

3. Transforming the complex into one that can pull away from the surface.

4. Complex is released from the mineral surface.

5. Complex diffusion into solution.

Among these five steps, the rate-limiting step can be either step 3 or 5 [17]. Chelating agents are frequently utilized at pH 4 in oilfield acidizing processes. This pH is adjusted by incorporating diluted HCl. It should be noted that despite the presence of H^+ at this pH, chelating agent assisted dissolution accounts for 90% of dissolution [18] via the previously stated surface complexation process. Surface complexation refers to the creation of a chelating agent-cation complex with cations on the surface. This strengthens the bond between the chelating agent and the ion while weakening the bonds of the crystal structure surrounding the ion. Due to the protonation of oxygen atoms in the presence of H^+ at this pH, oxide bonds in the lattice are also weakened. This allows the chelating agent to extract the atom, thus making surface complexation the major process of dissolution at low pH.

2. APCAs for mineral acidizing

2.1 Subsurface formations containing carbonate minerals

The primary objective of carbonate acidizing is to increase permeability through the creation of wormholes that bypass the formation damage close to the wellbore (**Figure 5**). Conventional matrix acidizing involves the use of strong acids such as HCl

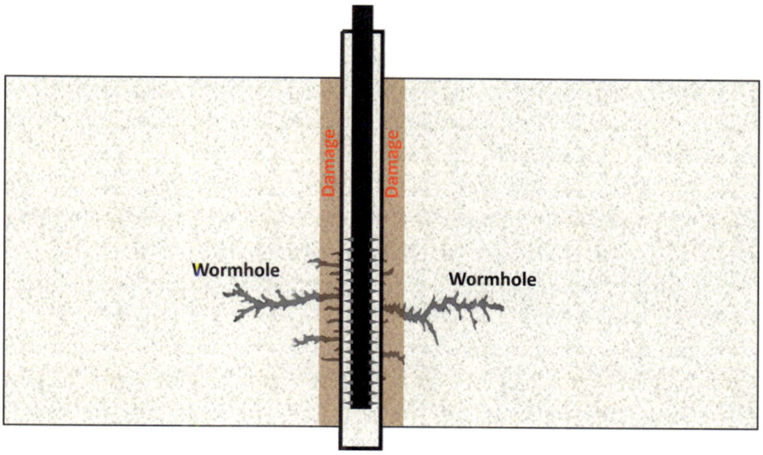

Figure 5.
Representation of a carbonate acidizing operation yielding wormholes that bypass the damage zone and improves conductivity of the formation.

which result in the consumption of large volumes of acid throughout the process [19]. However, at temperatures above 200°F (93°C), HCl becomes highly corrosive and causes face dissolution instead of forming wormholes. Therefore, chelating agents can be used as an alternative acidizing fluid in these conditions.

As shown earlier, chelating agents do not depend solely on hydrogen ions for their dissolution process, which is different from that of HCl. With chelating agents, chelation and hydrogen ion attack simultaneously at low pH, increasing dissolution [13, 18]. HCl and organic acids (e.g. formic and acetic acid) have a 2:1 stoichiometry, compared to 1:1 for regularly used chelating agents (acid:carbonate) [20]. The dissolution of calcium carbonate at pH between 4 and 5 by chelating agents can be expressed by Eqs. (3)-(5) [21]:

Attack with hydrogen:

$$2H^+ + CaCO_3 \rightarrow Ca^{2+} + CO_2 + H_2O. \tag{3}$$

Calcium chelation:

$$Ca^{2+} + H_2Y \rightleftarrows CaY^{2-} + 2H^+. \tag{4}$$

Mixture of both:

$$H_2Y^{2-} + CaCO_3 \rightleftarrows CaY^{2-} + H_2O + CO_2. \tag{5}$$

The use of chelating agents to acidize carbonate formations has been extensively studied by many authors. The first to do so were Fredd and Fogler [13], who showed that wormholes could be created by injecting chelating agents at pH ranges of 4–13 and noted the absence of sludge or asphaltene formation throughout the process. They later investigated the kinetics of carbonate dissolution using CDTA, DTPA, and EDTA at pH ranges of 3–12 [18]. In this work, they determined that the pH, type, and concentration of the chelating agent were all factors in the dissolution.

Subsequently, many authors have investigated EDTA as an alternative acidizing fluid. Due to its lower reactivity than HCl [21], EDTA was observed to produce more effective wormholes and could be used at temperatures up to 400°F (204°C) [22, 23].

Due to the poor solubility of EDTA, more soluble APCAs such as HEDTA and HEIDA were investigated by other authors. While not as effective, they were also shown to be able to produce dominant wormholes at temperatures between 250°F (121°C) and 400°F (204°C) [6, 20, 22, 24].

A more recent APCA developed was GLDA which overcomes the solubility issues with EDTA and is also biodegradable. This APCA was tested by Mahmoud et al. and was found to be effective at forming wormholes between 180°F (82°C) and 300°F (149°C) [23]. In addition, LePage et al. [25]. showed that GLDA had greater solubility and calcium dissolution capacity at 300°F (149°C) across a wide pH range.

In addition, the method of preparing chelating agents was also found to impact its performance in acidizing treatments. Chelating agent solutions are typically prepared by mixing the salt of the chelating agent with water and acidifying the resulting solution using conventional acids. Tests were done to examine the impact of using various acids (to get the chelating agent's pH to a value of 4) on the carbonate acidizing performance [26]. The authors found that 5 wt.% GLDA adjusted with acetic acid dissolved limestone with greater efficacy than the same solution adjusted with HCl.

2.2 Subsurface formations containing sandstone minerals

Unlike carbonate acidizing, acid treatments in sandstone reservoirs are not aimed towards increasing the native matrix permeability, but rather to repair the damage that was done during drilling and completion operations. Clay, carbonates, sulfates, zeolite, feldspar, sand, and oxides are just a few of the minerals found in sandstone reservoirs. While clays are present in both carbonate and sandstone formations, acidizing in sandstone reservoirs can become complicated due to the presence of cementing materials. These cementing materials are often limestone and act to hold individual sand grains together. If they are removed during treatment, sand production can occur, resulting in damage to pipes and equipment due to erosion [27, 28]. Therefore, using a weaker stimulation fluid such as a chelating agent instead of HCl, helps to reduce the possibility of sand production.

Furthermore, when selecting chelating agents for sandstone acidizing, the type of counter-cation must also be considered. Commercial chelating agents are produced in a variety of salts with cations such as Na^+, K^+, or NH_4^+. Ammonium salts of chelating agents should be utilized in hydrofluoric (HF) acid applications to avoid fluorosilicate precipitation. While this is true for common chelating agents, recent research has shown that chelating agents can be designed to overcome the problem of fluorosilicate precipitation with non-ammonium chelating agents [29, 30].

Both laboratory and field treatments have demonstrated the utility of common APCAs for stimulating sandstone. It has been shown that chelating agents can sequester cations such calcium, iron, magnesium, and aluminum, which inhibits various precipitations from occurring during the process of sandstone acidizing [31–34]. In laboratory tests, sandstone cores were subjected to a variety of common chelating agent treatments at approximately 200°F (93°C) and permeability was found to have increased [35, 36]. In these experiments, EDTA was found to cause formation damage by precipitating in the core and clogging the pores due to its poor solubility. It was also discovered that although chelating agents helped to increase the permeability of the core, the HF main flush decreased the permeability due to secondary formation damage. A new chelating agent treatment combined with HF was evaluated by Mohd Jeffry et al. [37] in sandstone cores. While the formulation of the chelating agent was not shown, the authors observed a 264% increase in permeability after two such treatments were used.

Following the successful laboratory experiments, chelating agents have therefore been applied to field treatments and have been shown to be able to restore permeability to damaged formations [29, 30, 36]. However, authors have described different methods of applying chelating agent treatments. Legemah et al. [38] described a two-step process, using a low volume, high concentration formulation for the first treatment and a high volume, low concentration formulation for the second. Typically, the chelating agent treatment is applied in a single step. Therefore, before field application of a chelating agent stimulation treatment, it is necessary to conduct laboratory scale tests to determine the method of treatment as well as to identify other potential issues that can arise, especially with sandstone reservoirs.

3. Chelation of iron

The chelating properties of chelating agents can be used to tackle another major problem in the oil and gas industry, namely iron. Ferric (Fe^{3+}) iron has a reputation for causing many issues during stimulation treatments by stabilizing damaging emulsions and inducing the precipitation of sludge and asphaltene [39, 40]. Iron has also been found to be incompatible with several additives used in the oil industry, including viscoelastic surfactants (VES). In addition, it is well known that iron hydroxides easily precipitate, which damages the formation by clogging the pores and lowering permeability. Iron fluoride precipitation is an additional issue when HF is used in sandstone acidizing. Taylor et al. [41, 42] studied the precipitation of iron hydroxide at room temperature, at spent HCl acid conditions, and in the absence of gases such as H_2S. They observed that ferric ions (Fe^{3+}) precipitate at pH above 1, while ferrous ions (Fe^{2+}) precipitate at a pH above 6 as shown in Eqs. (6) and (7):

When pH > 1:

$$Fe^{3+} + 3OH^- \rightarrow Fe(OH)_3.mH_2O. \qquad (6)$$

When pH > 6:

$$Fe^{2+} + 2OH^- \rightarrow Fe(OH)_2.mH_2O. \qquad (7)$$

Iron can be produced from a number of sources in the formation, such as formation rock and tubulars (Fe^{2+}, Fe^{3+}), formation water (Fe^{2+}), and treatment fluids (Fe^{3+}). The concentration of iron in fluids during an acid treatment may reach 4000 mg/L at the wellhead and may surpass 100,000 mg/L by the time it reaches the formation [43, 44]. As a result, competent iron control is needed to prevent precipitation, incompatibilities, and formation damage.

Spent acids typically have a pH between 4 and 4.5 due to the formation of carbonic acid when carbon dioxide, a product of carbonate acidizing, dissolves in water. As a result, any Fe^{3+} ions present will precipitate as Fe^{3+} hydroxide when the pH reaches this level. To prevent this, iron-control agents such as APCAs are added to chelate the ions and keep them in solution. One of the first and most used iron control agents was citric acid, as illustrated by Hall and Dill [45]. However, it loses 50% of its potency at 350°F (176°C) and can result in the precipitation of calcium citrate and asphaltene/sludge issues [46].

The effectiveness of chelating agents as iron control agents is dependent on its solubility and its iron stability constants. If the chelating agent experiences solubility issues, such as EDTA, in the acidizing fluid, then it cannot chelate iron ions. Furthermore, the stability constant of the iron-chelating agent complex is also important due to competition from hydroxide ions in the solution. Therefore, the Fe^{3+}-chelating agent complex must be more stable than Fe^{3+} hydroxide to prevent the precipitation of Fe^{3+} hydroxide [42]. In general, chelating agents have greater stability constants for iron and can keep iron soluble in solutions over a wide pH range as shown in **Figure 6**.

It should be noted that in sour environments (H_2S), iron control is more difficult as hydrogen sulfide is a strong reducing agent that causes precipitation through the formation of iron sulfides of low solubility and higher stability compared to the iron-chelating agent complex. As certain forms of iron sulfide can be hard to remove, it is ideal to prevent their formation instead of remediation. Consequently, hydrogen sulfide scavengers must be used in combination with chelating agents to keep iron soluble in solution in sour environments [48, 49]. More information regarding iron sulfide scale is provided later in this chapter.

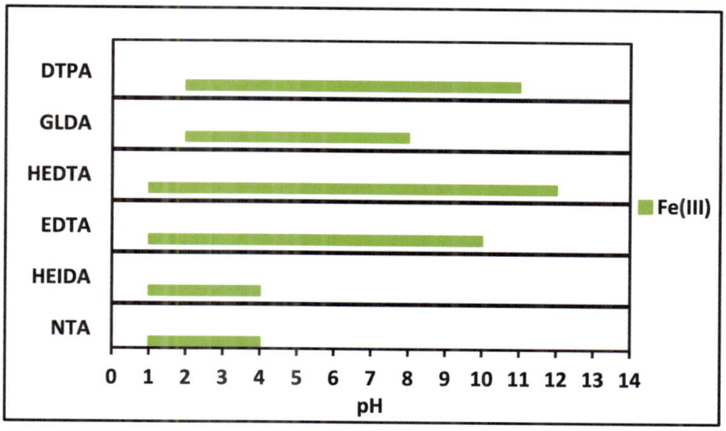

Figure 6.
Chelating agent optimum pH range for iron (III) (re-graphed from [47]).

4. Inorganic scale removal

4.1 Calcium and barium sulfate

Inorganic scale is a huge problem in the oilfield as it tends to form restrictions in the wellbore and the surface pipelines that ultimately result in reducing production rates. Inorganic scale precipitation results from mixing two incompatible water sources, where the first contains an abundance of metal ions such as calcium, magnesium, or barium, and the other contains anions such as bicarbonates or sulfates. Another form of inorganic scale precipitation can occur through self-precipitation initiating from one water source. This occurs due to changes in the temperature and pressure conditions which ultimately change the thermodynamic equilibrium of ions in solution causing it to form inorganic scales. Inorganic scale precipitation is seen at choke vales, bends in the surface pipelines, the top 500 ft of the well, or even downhole at the intake of electrical submersible pumps.

The two most common types of inorganic scales faced in the oilfield are carbonate and sulfate precipitations. While inorganic acids such as HCl provide a cheap and easy method to remove carbonate scales, they are often unable to dissolve sulfate scales. In this scenario, chelating agents are required [48].

In the past, EDTA was commonly used to treat calcium scales in boilers and oilfield surface pipelines [49–51]. Following that, MGDA and HEIDA were tested and proved to be effective [20, 52, 53]. It was also shown that GLDA could work for this function as well [54].

Barium sulfate ($BaSO_4$) is the hardest to remove due to its low solubility in water (2 mg/L) [55, 56]. Chelating agents such as EDTA, HEDTA, and DTPA were extensively studied in barium sulfate scale removal treatments [57–59]. After recognizing the difficulty of dissolving barium sulfate in the oil and gas industry, authors tested and compared several chelating agents (DTPA, DOTA, CDTA and EDTA) for removal treatments [60]. Through the testing it was noted that DTPA at pH above 12 gave the best results, even when compared to chelating agents that have higher complex stability constants with barium (Ba-DOTA > Ba-DTPA).

More recently, researchers showed that larger chelating agents such as macropa and macroquin were able to outperform DTPA and DOTA at pH 8 and pH 11 conditions [61]. This opened an area of research to compare these bigger molecules to common oilfield chelating agents. Through these studies authors were able to show that the use of macropa at pH 8 outperformed the use of EDTA, HEDTA, or DTPA. Although, at high pH (>12), a dilute solution of penta potassium DTPA showed the most prominent results in barium sulfate dissolution [56, 62–64].

It was also shown that combinations of chelating agents such as DTPA/GLDA, DTPA/EDTA, or DTPA/MGDA resulted in lower performance than DTPA alone. This is mainly because barite dissolution is inhibited by the physical presence of other chelating agent molecules [65, 66]. It was also noted that barite dissolution with DTPA increases as temperature increased, indicating a thermally activated process [62, 67, 68].

Furthermore, researchers have also shown that the performance of some chelating agents decreases with an increase in concentration due to steric hindrance of chelating agent molecules on the surface of the scale [69]. Consequently, it is recommended to apply a multi-cyclic injection technique with dilute chelating agent solution [62, 70].

Several chelating agent and dissolution catalyst formulations have been proposed for sulfate scale dissolution [56, 71, 72]. Examples of dissolution catalysts are oxalic acid, fluoride, dithionate, citrate, thiosulfate, nitriloacetate, mercaptoacetate, hydroxyacetate, amino acetate and formate brine [56, 62, 71, 73–79]. Carbonate and formate ions are commonly used today [62, 71, 80]. Halogens such as fluoride show very good synergy with EDTA but negatively affect DTPA's ability to dissolve barium sulfate [56, 67].

Paul and Fieler [56] showed the importance of cation choice to achieve a successful scale solvent, especially when a sizable amount is added to adjust the pH. Their testing compared lithium, sodium, potassium, and cesium and pointed out that the dissolution is enhanced as the size of the cation is increased. Of these cations, potassium would be the optimum choice since cesium is difficult to obtain.

4.2 Iron sulfide scale

Another inorganic scale seen in the oilfield is sulfide-based scales. Although zinc, lead, and mercury sulfide do exist in the oilfield, iron sulfide is the most common sulfide scale. This is due to the abundance of iron sources in the oilfield. Iron sulfide is black colored, oil wet, heterogenous, and explosive scale. It can come in many different forms depending on a variety of factors including temperature, pressure, and exposure time to H_2S.

Iron sulfide scale can be dissolved by conventional acid systems if the iron to sulfur ratio is close to unity [81, 82]. However, this reaction can regenerate (H_2S) and precipitates elemental sulfur. Therefore, H_2S scavengers must be added if HCl or chelating agents are to be used [83–85]. In cases where the iron to sulfur ratios is close to 0.5, iron sulfide removal can become difficult to dissolve through conventional systems. This is a result of the iron being protected by sulfur inside the mineral structure, which prevents the iron from being leached by acids, making its dissolution difficult. Common solutions in this scenario include milling or using oxidizing agents such as chlorites/chlorine dioxide or permanganates to oxidize iron sulfide to soluble Fe^{3+} ions [86, 87]. **Table 2** shows a list of common iron sulfide species.

Research has shown that EDTA at pH 8–10 was capable of dissolving almost 80 wt.% of iron sulfide scale after 20 hours at 175°F (80°C) [88]. Additionally, the use of 20 wt.% EDTA and DTPA were also determined to be capable of preventing iron

Iron sulfide	Chemical composition	Iron:sulfur ratio
Mackinawite	Fe_9S_8	~1
Troilite	FeS	1
Pyrrhotite	Fe_7S_8	~1
Pyrite	FeS_2	0.5
Marcasite	FeS_2	0.5

Table 2.
Common iron sulfide scale [71].

sulfide formation with an inhibition efficiency of 83 wt.% and 86 wt.%, respectively [89]. Other chelating agents such as GLDA and MGDA can also be used to remove FeS scale [78]. Furthermore, the use of converting agents such as (K_2CO_3) with DTPA at pH >11 proved to be efficient in dissolving iron sulfide scales with high concentration of sulfur [90]. The converting agent converts iron sulfide to iron carbonate which can be dissolved by chelating agents. This method can dissolve over 85 wt.% of the iron sulfide field scale samples at 155°F (68°C).

5. Steel corrosion by aminopolycarboxylic acids

Considering chelating agents are frequently used as substitutes for the conventional HCl acidizing recipes at high temperatures, it becomes essential to understand the capabilities of metal corrosion by chelating agent [10]. Corrosion to the tubulars can lead to a variety of concerns regarding the well's safety. Moreover, corrosion can cause destruction to downhole pumps and equipment resulting in reduced production rates and significant financial losses.

The metal corrosion mechanism is comparable to that of mineral surfaces [91]. Through a process known as surface complexation, ligands attach and dissolve the metal oxide layer of the tubular.

As stated by Ng et al. [92], chelating compounds corrode at pH 4 through a two-step process that involves chelating agents that promote the removal of the iron oxide layer and a redox interaction between the chelating agent and the base metal. As the base metal Fe is being oxidized to Fe^{2+}, the chelating agent's carboxylic groups are being reduced into aldehydes.

Orthophosphates, silicates, chromates, amines, aldehydes, alkaloids, nitro and nitroso, thiourea, phenols, naphthol, and chelating chemicals are a few examples of corrosion inhibitors that can be utilized to decrease corrosion rates in challenging environments. The type and extent of corrosion, length of protection, and temperature affect the corrosion inhibitor choice.

The corrosion rates of chelating agents in low pH have been investigated as part of several tests using corrosion-resistant alloys and carbon steels. If the corrosion rates are kept under 0.05 lb/ft^2 over 6 h (45.4 mmpy), they are generally regarded as acceptable. For chrome steel, corrosion limit is lowered to 0.03 lb/ft^2 over 6 hours (27.2 mmpy) due to the higher material costs [93, 94]. The corrosion rate depends on many factors such as the APCA and the type of metal corroded. Some examples from literature tests at pH 4 can be seen in **Table 3** for low carbon steel tubulars, and in **Table 4** for corrosion resistant alloys.

Metal	APCA	Conc, wt.%	Temperature, °F (°C)	Inhibitor, v. %	Duration, hours	Corrosion rate, lb/ft^2 (mmpy)	Source
N-80	GLDA	20	300 (149)	0	12	0.724 (328.8)	[92]
	HEDTA	20	300 (149)	0	12	0.803 (364.6)	
	EDTA	20	300 (149)	0	12	0.858 (389.6)	
	MGDA	20	300 (149)	0	12	0.642 (291.5)	
	GLDA	20	350 (176)	0	12	0.754 (342.4)	
	HEDTA	20	350 (176)	0	12	0.974 (442.3)	
	EDTA	20	350 (176)	0	12	1.07 (485.9)	
	MGDA	20	350 (176)	0	12	0.76 (345.1)	
	HEDTA	20	350 (176)	1	12	0.0102 (4.6)	
	MGDA	20	350 (176)	1	12	0.00561 (2.5)	
L-80	GLDA	20	300 (149)	0	6	0.5937 (539.2)	[95]
	GLDA	20	300 (149)	0.005	6	0.0262 (23.8)	
	GLDA	20	300 (149)	0.001	6	0.0394 (35.8)	[96]
	GLDA	20	300 (149)	0	6	0.1956 (177.6)	
	GLDA	20	300 (149)	0.001	6	<0.05 (<45.4)	[97]
	HEDTA	20	300 (149)	0	6	0.8341 (757.5)	[95]
	HEDTA	20	300 (149)	0.005	6	0.1300 (118.1)	
	HEIDA	20	300 (149)	0	6	0.6519 (592.0)	[96]
	MGDA	20	300 (149)	0	6	0.4596 (417.4)	

Table 3.
Low-carbon steel (LCS) corrosion rate summary.

Metal	APCA	Conc., wt.%	Temperature, °F (°C)	Inhibitor, v. %	Duration, hours	Corrosion rate, lb/ft^2 (mmpy)	Source
13Cr	GLDA	20	300 (149)	0	6	0.0009 (0.8)	[96]
	GLDA	20	300 (149)	0	6	0.0080 (7.3)	[54]
	HEDTA	20	300 (149)	0	6	0.3310 (300.6)	
	HEDTA	20	300 (149)	0	6	0.5253 (477.1)	[98]
	HEIDA	20	300 (149)	0	6	0.0590 (53.6)	[96]
	MGDA	20	300 (149)	0	6	0.0900 (81.7)	
	ASDA	20	300 (149)	0	6	0.0563 (51.1)	
	HEDTA	20	350 (176)	0.3	4	0.0100 (9.1)	[91]
	GLDA	20	350 (176)	0.5	6	0.0496 (45.0)	[98]
	GLDA	20	350 (176)	0	6	0.3478 (315.9)	
S13Cr	GLDA	20	350 (176)	0	6	0.0187 (17.0)	[98]
22Cr	GLDA	20	300 (149)	0	6	0.0001 (0.09)	[99]

Table 4.
Corrosion resistant alloy (CRA) corrosion rate summary.

6. Degradation of aminopolycarboxylic acids

The effectiveness of chemical treatments in the oil and gas industry depends on the performance of the chemical agents. Its degradation under extreme conditions can render the treatment ineffective. In the oilfield, this can occur due to high bottomhole temperatures that cause the APCA to break down. Therefore, care must be taken when selecting the APCA for the treatment.

Thermal decomposition of chelating agents has been studied by many authors. Research on the degradation of NTA and EDTA has shown that EDTA hydrolyzes to produce HEIDA and IDA, around 500°F (260°C) and pH 9.5 [100]. HEIDA was later shown to undergo further hydrolysis to yield ethylene glycol and IDA.

N-methyliminodiacetic acid (MIDA), the first breakdown product of NTA, was converted at 560°F (293°C) into methylsarcosine, which was then converted into trimethylamine. This process was non-hydrolytic and proceeded stepwise, eventually yielding carbon monoxide, carbon dioxide, formaldehyde, and methylamines [101]. As a result of the average field temperatures being much lower, these decompositions are unlikely.

Several authors [20, 75–77, 102] investigated how GLDA decomposed thermally. When heated at 300°F (149°C) and 350°F (176°C), GLDA displayed thermal stability temperature equivalent to that of HEDTA; the breakdown products were cyclic GLDA and formic acid [25]. IDA, oxotetrahydrofuran-2 carboxylic acid, hydroxyglutaric acid, and acetic acid were identified as the GLDA products of decomposition through mass spectrometry [103].

At temperatures as high as 350°F (176°C), MGDA and HEIDA exhibited outstanding stability, showing no degradation after 6 hours [35]. But it was observed that ASDA decayed rapidly, with less than 10 wt.% of the APCA left after the same amount of time at 300°F (149°C). Since the byproducts of degradation will no longer be able to carry out the original ligand's intended function, such as controlling iron, removing scale, or acidifying the matrix, these results necessitate the application of an APCA with sufficient thermal stability for high-temperature wells encountered in the oilfield.

The biodegradability of APCA used in the oilfield has also come under close examination because of a recent rise in environmental regulations. Chelating agents in the environment have the power to chelate metal ions, which allows them to affect metal speciation and bioavailability as well as remobilize toxic heavy metals into groundwater [104–106]. Because of the potential for groundwater contamination, this adds another degree of consideration which must go into choosing the best APCA for treatment. Depending on the amount of nitrogen atoms that exist in the structure, chelating compounds can differ in their susceptibility to biodegradation. When compared to chelating compounds with numerous nitrogen atoms, those with a single nitrogen atom tend to be more biodegradable [107–110].

7. Summary

There are many challenges with applying conventional HF or HCl acids for oilfield stimulation. Chelating agent research has progressed over the previous decades to identify the best application for them in the petroleum industry as a replacement to conventional acid systems. Common chelating agents were addressed in this work, accompanied by a thorough analysis of the unique features that distinguish presented chelating agents. A summary of the structural formulas and chemical properties of some of the most common chelating agents was also provided.

Chelating agents have been utilized for oilfield stimulation either as a stand-alone treatment or in combination with traditional acids to lessen some of the downsides. They also show good performance even at high temperatures and serve to reduce the number of additives required for stimulation and scale removal treatments, such as corrosion inhibitors due to their low corrosivity. In addition, their ability to chelate iron is invaluable to prevent formation damage from iron hydroxide precipitation and sludge formation, common phenomena that can quickly result in severe formation damage. Furthermore, many chelating agents are environmentally friendly and will decompose readily when exposed to the elements and minimize environmental impact. Despite this, they are still resistant to thermal decomposition and are able to perform high temperature treatments.

Despite these benefits, it is important to exercise care when selecting the appropriate chelating agent for the application. Due to the wide variety of options, properties such as solubility, stability constants, reactivity, and price must be considered when choosing the most appropriate chelating agent. By providing a comprehensive summary of the chemistry and applications of chelating agents in the oil and gas industry, we hope readers will be able to select the best chelating agent for their treatment.

Nomenclature

APCA	aminopolycarboxylic acids
ASDA	L-aspartic acid N, N-diacetic acid
CDTA	trans-l,2-cyclohexylenediaminetetraacetic acid
CRA	corrosion resistant alloys
DOTA	1,4,7,10-tetraazacyclododecane-1,4,7,10-tetraacetic acid
DTPA	diethylenetriaminepentaacetic acid
EDTA	ethylenediaminetetraacetic acid
GLDA	L-glutamic acid N, N-diacetic acid
HCl	hydrochloric acid
HEDTA	hydroxyethyl ethylenediaminetriacetic acid
HEIDA	hydroxyethyliminodiacetic acid
HF	hydrofluoric acid
IDA	iminodiacetic acid
LCS	low carbon steel
MGDA	methylglycinediacetic acid
MIDA	N-methyliminodiacetic acid
NTA	nitrilotriacetic acid
VES	viscoelastic surfactant

Author details

Tariq Almubarak[1]* and Clarence Ng[2]

1 Saudi Aramco EXPEC ARC, Dhahran, Saudi Arabia

2 Independent Researcher, Naperville, USA

*Address all correspondence to: tariq.mubarak@hotmail.com

References

[1] Almubarak T, Ng JH, Ramanathan R, Nasr-El-Din HA. Chelating agents for oilfield stimulation: Lessons learned and future outlook. Journal of Petroleum Science and Engineering. 2021;**205**:108832-108832. DOI: 10.1016/j.petrol.2021.108832

[2] Dwyer FP, Mellor DP. Chelating Agents and Metal Chelates. London: Academic Press; 1964

[3] Martell AE, Calvin M. Chemistry of the Metal Chelate Compounds. New York: Prentice-Hall Inc; 1952

[4] Schwarzenbach G, Ackermann H. Komplexone XII. Die Homologen der Äthylendiamin-tetraessigsäure und ihre Erdalkalikomplexe. Helvetica Chimica Acta. 1948;**31**(4):1029-1048. DOI: 10.1002/hlca.19480310407

[5] Begum ZA, Rahman IMM, Sawai H, Tate Y, Maki T, Hasegawa H. Stability constants of Fe(III) and Cr(III) complexes with dl-2-(2-carboxymethyl) nitrilotriacetic acid (GLDA) and 3-Hydroxy-2,2′-iminodisuccinic acid (HIDS) in aqueous solution. Journal of Chemical & Engineering Data. 2012;**57**(10):2723-2732. DOI: 10.1021/je300593

[6] Frenier WW, Fredd CN, Chang F. Hydroxyaminocarboxylic acids produce superior formulations for matrix stimulation of carbonates. In: Presented at the SPE European Formation Damage Conference, The Hague, Netherlands. 21-22 May 2001. SPE-68924-MS. DOI: 10.2118/68924-MS

[7] Martell AE, Smith RM, Motekaitis RJ. NIST Standard Reference Database 46: NIST Critically Selected Stability Constants of Metal Complexes Database (Version 8.0 For Windows). College Station, TX: Texas A&M University; 2004

[8] Begum ZA, Rahman IMM, Tate Y, Sawai H, Maki T, Hasegawa H. Remediation of toxic metal contaminated soil by washing with biodegradable aminopolycarboxylate chelants. Chemosphere. 2012;**87**(10):1161-1170. DOI: 10.1016/j.chemosphere.2012.02.032

[9] Luyster M, Patel A, Ali SA. Development of a delayed-chelating cleanup technique for openhole Gravel-Pack horizontal completions using a reversible invert emulsion drill-in system. In: Presented at the SPE International Symposium and Exhibition on Formation Damage Control, Lafayette, Louisiana, USA. 15-17 February. SPE-98242-MS. DOI: 10.2118/98242-MS

[10] Ramanathan R, Nasr-El-Din HA, Zakaria A. New insights into the dissolution of iron sulfide using chelating agents. Spe Journal. 2020;**25**(06):3145-3159. DOI: 10.2118/202469-PA

[11] Almubarak T, Ng JH, Nasr-El-Din H. Oilfield Scale Removal by Chelating Agents: An Aminopolycarboxylic Acids Review. Day 4 Wed, April 26, 2017. 2017. DOI: 10.2118/185636-MS

[12] Ramanathan R, Nasr-El-Din HA. A comparative experimental study of alternative iron sulfide scale dissolvers in the presence of oilfield conditions and evaluation of new synergists to aminopolycarboxylic acids. SPE Journal. 2021;**26**(02):693-715. DOI: 10.2118/205005-PA

[13] Fredd CN, Fogler HS. Chelating Agents as Effective Matrix Stimulation Fluids for Carbonate Formations. In: Presented at

the International Symposium on Oilfield Chemistry, Houston, Texas. 18-21 Feb SPE-37212-MS. DOI: 10.2118/37212-MS

[14] Chang HC, Matijević E. Interactions of metal hydrous oxides with chelating agents. Journal of Colloid and Interface Science. 1983;**92**(2):479-488. DOI: 10.1016/0021-9797(83)90169-8

[15] Carbonaro RF, Gray BN, Whitehead CF, Stone AT. Carboxylate-containing chelating agent interactions with amorphous chromium hydroxide: Adsorption and dissolution. Geochimica et Cosmochimica Acta. 2008;**72**(13):3241-3257. DOI: 10.1016/j.gca.2008.04.010

[16] Ramanathan R, Nasr-El-Din HA. Improving the Dissolution of Iron Sulfide by Blending Chelating Agents and its Synergists. In: Paper presented at the SPE Middle East Oil and Gas Show and Conference, Manama, Bahrain. Mar 2019. DOI: 10.2118/195128-MS

[17] Nowack B, Sigg L. Dissolution of Fe(III) (hydr) oxides by metal-EDTA complexes. Geochimica et Cosmochimica Acta. 1997;**61**(5):951-963. DOI: 10.1016/S0016-7037(96)00391-2

[18] Fredd CN, Fogler HS. The influence of chelating agents on the kinetics of calcite dissolution. Journal of Colloid and Interface Science. 1998;**204**(1):187-197. DOI: 10.1006/jcis.1998.5535

[19] Kotb A, Nasr-El-Din HA. New insights into mass transfer when using the rotating disk apparatus for Newtonian and non-Newtonian fluids. SPE Journal. 2021;**26**(03):1161-1173. DOI: 10.2118/201246-PA

[20] Frenier WW, Fredd CN, Chang F. Hydroxyaminocarboxylic acids Produce superior formulations for matrix stimulation of carbonates at high temperatures. In: Presented at the SPE Annual Technical Conference and Exhibition, New Orleans, Louisiana. 30 Sept-3 Oct. SPE-71696-MS. DOI: 10.2118/71696-MS

[21] Shaughnessy CM, Kline WE. EDTA removes formation damage at Prudhoe Bay. Journal of Petroleum Technology. 1983;**35**(10):1783-1791. DOI: 10.2118/11188-PA

[22] Barri A, Hassan A, Mahmoud M. Carbonate stimulation using chelating agents: Improving the treatment performance by optimizing the fluid properties. ACS Omega. 2022;**7**(10):8938-8949. DOI: 10.1021/acsomega.1c07329

[23] Mahmoud M, Nasr-El-Din HA, De Wolf C, LePage JN. An effective stimulation fluid for deep carbonate reservoirs: A core flood study. In: Paper presented at the International Oil and Gas Conference and Exhibition in China, Beijing, China. June 2010. DOI: 10.2118/131626-MS

[24] Huang T, McElfresh P, Gabrysch A. Carbonate matrix acidizing fluids at high temperatures: Acetic acid, chelating agents or long-chained carboxylic acids? In: Paper presented at the SPE European Formation Damage Conference, The Hague, Netherlands. May 2003. DOI: 10.2118/82268-MS

[25] LePage JN, De Wolf C, Bemelaar J, Nasr-El-Din HA. An environmentally friendly stimulation fluid for high-temperature applications. Spe Journal. 2011;**16**(01):104-110. DOI: 10.2118/121709-PA

[26] Shi Y, Yu L, Chen S, He Y, Yang X, Duan L, et al. Effects of L-glutamic acid, N, N-diacetic acid as chelating agent on acidification of carbonate reservoirs in acidic environments. Journal of Natural Gas Science and Engineering. 2020;**82**:103494-103494

[27] Amaefule JO, Kersey DG, Norman DI, Shannon PM. Advances In formation damage assessment and control strategies. Annual Technical Meeting. In: Paper presented at the Annual Technical Meeting, Calgary, Alberta. June 1988. DOI: 10.2118/88-39-65

[28] Civan F. Reservoir Formation Damage: Fundamentals, Modeling, Assessment, and Mitigation. Houston, Tex.: Gulf Pub. Co; 2000

[29] Reyes EA, Smith AL, Beuterbaugh A. Properties and applications of an alternative Aminopolycarboxylic acid for acidizing of sandstones and carbonates. In: Paper presented at the SPE European Formation Damage Conference & Exhibition, Noordwijk, The Netherlands. June 2013. DOI: 10.2118/165142-MS

[30] Smith AL, Woon GC, Smits F, Govinathan K, Yong WJ, Beuterbaugh A, et al. Field Results and Experimental Comparative Analysis of Sodium and Nonsodium Chelant-Based HF Acidizing Fluids for Sand Control Operations. In: Paper presented at the SPE International Conference and Exhibition on Formation Damage Control, Lafayette, Louisiana, USA. Feb 2016. DOI: 10.2118/179001-MS

[31] Frenier WW, Fu D, Davies SN. Method for Single-Stage Treatment of Siliceous Subterranean Formations. US Patent Application No. 20,090,192,057. 2009

[32] Frenier WW, Ziauddin M, Davies S. Composition Comprising a Fully Dissolved Non-HF Fluoride Source and Method for Treating a Subterranean Formation. U.S. Patent No. 7,589,050. 2009

[33] Frenier WW, Fu D, Davies SN. Method for Single-Stage Treatment of Siliceous Subterranean Formations. U.S. Patent No. 8,316,941. 2012

[34] Wei L, Qing W, Changfeng X, Mengchuan Z, Xinmin K, Erdong Y, et al. A green and environmentally friendly chelated retarding acid for acidification of sandstone reservoirs. In: Proceedings of the ASME 2022 41st International Conference on Ocean, Offshore and Arctic Engineering. Volume 10: Petroleum Technology. Hamburg, Germany. 5-10 June 2022. V010T11A028. ASME. DOI: 10.1115/OMAE2022-79271

[35] Gomaa I, Mahmoud M. Sandstone acidizing using a low-reaction acid system. Journal of Energy Resources Technology-transactions of the Asme. 2020;**142**(10):103008-1 to 103008-7. DOI: 10.1115/1.4047317

[36] Shafiq MU, Ben Mahmud HK, Arif M. Mineralogy and pore topology analysis during matrix acidizing of tight sandstone and dolomite formations using chelating agents. Journal of Petroleum Science and Engineering. 2018;**167**:869-876. DOI: 10.1016/j.petrol.2018.02.057

[37] Mohd Jeffry SJ, Mohd R, Mohd A, Madon B, Mohamad KI, Adlan E. Breakthrough formulation of novel modified HF acid system: Realizing remarkable value in sandstone reservoirs. In: Paper presented at the Abu Dhabi International Petroleum Exhibition & Conference, Abu Dhabi, UAE. Nov 2020. DOI: 10.2118/202895-MS

[38] Legemah MU, Gomaa A, Bilden D, Lowe C, Boles J, Qu Q, et al. Sequential injection process enhances acidizing treatment of high-temperature wells. In: Paper presented at the SPE Production and Operations Symposium, Oklahoma City, Oklahoma, USA. Mar 2015. DOI: SPE-173626-MS

[39] Jacobs IC, Thorne. Asphaltene precipitation during acid stimulation treatments. In: Paper presented at the SPE Formation Damage Control

Symposium, Lafayette. Louisiana. Feb 1986. DOI: 10.2118/14823-MS

[40] Almubarak T, AlKhaldi M, Ng JH, Nasr-El-Din HA. Matrix acidizing: A laboratory and field investigation of sludge formation and removal of oil-based drilling mud filter cake. SPE Drilling & Completion. 2020;**36**(02):281-299. DOI: 10.2118/178034-PA

[41] Taylor KC, Nasr-El-Din HA, Al-Alawi MJ. Systematic study of iron control chemicals used during well stimulation. SPE Journal. 1999;**4**(01):19-24. DOI: 10.2118/54602-PA

[42] Taylor KC, Nasr-El-Din HA. A systematic study of iron control chemicals - Part 2. In: Paper presented at the SPE International Symposium on Oilfield Chemistry, Houston, Texas, Feb 1999. DOI: 10.2118/50772-MS

[43] Gougler PD, Hendrick JE, Coulter AW. Field investigation identifies source and magnitude of iron problems. Paper presented at the SPE Production Operations Symposium, Oklahoma City, Oklahoma. Mar 1985. DOI: 10.2118/13812-MS

[44] Dill W, Smolarchuk P. Iron control In fracturing and acidizing operations. Journal of Canadian Petroleum Technology. 1988;**27**(03):75-78. DOI: 10.2118/88-03-08

[45] Hall BE, Dill WR. Iron control additives for limestone and sandstone acidizing of sweet and sour wells. In: Paper presented at the SPE Formation Damage Control Symposium, Bakersfield, California. Feb 1988. DOI: 10.2118/17157-MS

[46] Alkhaldi MH, Nasr-El-Din HA, Sarma H. Kinetics of the reaction of citric acid with calcite. SPE Journal. 2010;**15**(03):704-713. DOI: 10.2118/118724-PA

[47] Akzonobel. Chelate Product Guide. 2015. Available from: https://www.akzonobel.com/dissolvine/products/product_guide/ [Accessed: 01 September 2023]

[48] Clemmit AF, Ballance DC, Hunton AG. The dissolution of scales in oilfield systems. In: Offshore Europe, Aberdeen, United Kingdom. Sept 1985. DOI: 10.2118/14010-MS

[49] Bodine MW, Fernalld TH. Edta dissolution of gypsum, anhydrite, and Ca-Mg carbonates. SEPM Journal of Sedimentary Research. 1973;**43**:1152-1156. DOI: 10.1306/74D72932-2B21-11D7-8648000102C1865D

[50] Jamialahmadi M, Muller-steinhagen H. Reduction of calcium sulfate scale formation during nucleate boiling by addition of EDTA. Heat Transfer Engineering. 1991;**12**(4):19-26. DOI: 10.1080/01457639108939760

[51] Alotaibi F, Almubarak T, Alhamad L, Al MA. Chelating chemicals: A solution to end the battle with calcium sulfate scale. In: Paper presented at the Middle East Oil, Gas and Geosciences Show, Manama, Bahrain. Feb 2023. DOI: 10.2118/213502-MS

[52] Zack KL , Borst JP, DuRocher D, Przybyla DE, Leung V, Decker G. International Pat., WO2012142396. 2012

[53] Athey PS. Method for Controlling Alkaline Earth and Transition Metal Scaling with 2-Hydroxyethyliminodiacetic Acid. US Patent No. 5972868. 1999

[54] De Wolf CA. Ammonium salts of chelating agents and their use in oil and

gas field applications. International Patent Application WO/2012/080299. 2012

[55] Diebler H. Stability constants of metal-Ion complexes. Part A. Inorg Ligands. Vol. 21 aus: IUPAC Chemical Data Series. Pergamon Press, Oxford, New York, Toronto, Sydney, Paris, Frankfurt 1982. 310 Seiten. In: Högfeldt E. editor. Berichte der Bunsengesellschaft für physikalische Chemie. 1982;**87**:1227-1227. DOI: 10.1002/bbpc.19830871236

[56] Paul JM, Fieler ER. A new solvent for oilfield scales. In: Paper presented at the SPE Annual Technical Conference and Exhibition, Washington D.C., Oct 1992. DOI: 10.2118/24847-MS

[57] Rhudy JS. Removal of mineral scale from reservoir core by scale dissolver. In: Paper presented at the SPE International Symposium on Oilfield Chemistry, New Orleans, Louisiana. Mar 1993. DOI: 10.2118/25161-MS

[58] Kotlar HK, Selle OM, Haavind F. A "Standardized" method for ranking of scale dissolver efficiency. A Case Study from the Heidrun Field. In: Paper presented at the International Symposium on Oilfield Scale, Aberdeen, United Kingdom. Jan 2002. DOI: 10.2118/74668-MS

[59] Nasr-El-Din HA, Al-Mutairi SH, Al-Hajji HH, Lynn JD. Evaluation of a new barite dissolver: Lab studies. In: Paper presented at the SPE International Symposium and Exhibition on Formation Damage Control, Lafayette, Louisiana. Feb 2004. DOI: 10.2118/86501-MS

[60] Shuler PJ, Wang K-S, Dunn K. Effects of scale dissolvers on barium sulfate deposits: A macroscopic and microscopic study. In: Presented at CORROSION 2002, Denver, Colorado; April 2002. pp. 7-11

[61] Thiele NA, MacMillan SN, Wilson JJ. Rapid dissolution of BaSO4 by macropa, an 18-membered macrocycle with high affinity for Ba2+. Journal of the American Chemical Society. 2018;**140**(49):17071-17078. DOI: 10.1021/jacs.8b08704

[62] Putnis A, Putnis CV, Paul JM. The efficiency of a DTPA-Based solvent in the dissolution of barium sulfate scale deposits. In: Paper presented at the SPE International Symposium on Oilfield Chemistry, San Antonio, Texas. Feb 1995. DOI: 10.2118/29094-MS

[63] Blanco M, Tang Y, Shuler P, Goddard WA. Activated complex theory of barite scale control processes. Molecular Engineering. 1997;**7**(3/4):491-514. DOI: 10.1023/A:1008253202081

[64] Hassan A, Kamounah FS, Pittelkow M, Sølling TI, Mahmoud M. Barite removal using macropa: Simulation and solubility investigations. Energy & Fuels. 2022;**36**(16):9101-9110. DOI: 10.1021/acs.energyfuels.2c02038

[65] Shende AV. Dissolution of barite scale using chelating agents [MS thesis]. College Station, Texas: Texas A&M University; 2012

[66] Jaberi JA, Ahmed A, Bageri B, Elsayed M, Mahmoud M, Patil S, et al. Minimizing the barite scale in carbonate formations during the filter cake removal process. ACS Omega. 2022;**7**(21):17976-17983. DOI: 10.1021/acsomega.2c01339

[67] Dunn K, Yen TF. Dissolution of barium sulfate scale deposits by chelating agents. Environmental Science & Technology. 1996;**33**(16):2821-2824. DOI: 10.1021/es980968j

[68] Putnis CV, Kowacz M, Putnis A. The mechanism and kinetics

of DTPA-promoted dissolution of barite. Applied Geochemistry. 2008;**23**(9):2778-2788. DOI: 10.1016/j.apgeochem.2008.07.006

[69] Lakatos I, Lakatos-Szabo J. Potential of different polyamino carboxylic acids as barium and strontium sulfate dissolvers. In: Paper presented at the International Symposium and Exhibition on Formation Damage Control, Lafayette, Louisiana. Feb 2002. DOI: 10.2118/73719-MS

[70] Lakatos I, Lakatos-Szabo J, Kosztin B. Comparative study of different barite dissolvers: Technical and economic aspects. 2002. DOI: 10.2118/73719-MS

[71] Morris RL, Paul J. Method for Removing Alkaline Sulfate Scale US Patent No. 4980077. 1990

[72] Paul JM, Morris RL. Composition for Removing an Alkaline Earth Sulfate Scale. US Patent No. 5282995. 1994

[73] Morris RL, Paul JM. Method for Removing Alkaline Sulfate Scale. US Patent 5,049,297. 1991

[74] Morris RL, Paul JM. Sulfate Scale Dissolution. US Patent 5,084,105. 1991

[75] Morris R, Paul J. Composition for dissolution of sulfate scale. US Patent 5,259,980A. 1993

[76] Morris R, Paul J. Method for removing Alkaline Sulfate Scale. US Patent 4,980,077A. 1993

[77] Morris R, Paul J. Composition for removing an alkaline earth metal sulfate scale. US Patent 5,282,995A. 1994

[78] Muhala TF. Barium sulfate removal and anti-deposition compositions and process of use therefor. US Patent 4708805. 1991

[79] Howard S, Anderson Z, Parker SH. Solubility of barium sulfate in formate brines: New insight into solubility levels and reaction mechanisms. In: Paper presented at the SPE International Conference and Exhibition on Formation Damage Control, Lafayette, Louisiana, USA. Feb 2016. DOI: 10.2118/179021-MS

[80] Quattrini FJ. Aqueous reactive scale solvent. US patent 3660287A. 1970

[81] Ford W, Walker ML, Halterman MP, Parker DL, Brawley DG, Fulton R. Removing a typical iron sulfide scale: The scientific approach. 1992. DOI: 10.2118/24327-MS

[82] Nasr-El-Din HA, Al-Humaidan AY. Iron sulfide scale: Formation, removal and prevention. In: Paper presented at the International Symposium on Oilfield Scale, Aberdeen, United Kingdom. Jan 2001. DOI: 10.2118/68315-MS

[83] Frenier WW, Coffey MD, Huffines JD, Smith DC. Composition and Method for Removing sulfide containing Scale from Metal Surfaces. US Patent No. 4220550. 1980

[84] Crowe CW. Method of Preventing Precipitation of Ferrous Sulfide and Sulfur During Acidizing, US Patent No. 4633949. 1987

[85] Dill WR, Walker ML. Composition and Method for Controlling Precipitation When Acidizing Sour Wells, US Patent No. 4888121. 1989

[86] McCafferty JF, Tate EW, Williams DA. Field performance in the practical application of chlorine dioxide as a stimulation enhancement fluid. SPE Production & Facilities. 1993;**8**(01):9-14. DOI: 10.2118/20626-PA

[87] Romaine J, Strawser TG, Knippers ML. Application of chlorine

dioxide as an oilfield facilities treatment fluid. SPE Production & Facilities. 1996;**11**(01):18-21. DOI: 10.2118/29017-PA

[88] Martin LD. Methods and Compositions for Removing Deposits Containing Iron Sulfide from Surfaces Comprising Basic Aqueous Solutions of Particular Chelating Agents. US Patent No. 4276185. 1981

[89] Gasimli N, Mahmoud M, Patil S, Alsaiari HA, Hussein IA. Iron sulfide scale inhibition in carbonate reservoirs. ACS Omega. 2022;7(30):26137-26153. DOI: 10.1021/acsomega.2c01568

[90] Mahmoud M, Kamal MS, Bageri B, Hussein IA. Removal of pyrite and different types of iron sulfide scales in oil and gas wells without H2S generation. In: Paper presented at the International Petroleum Technology Conference, Doha, Qatar. Dec 2015. DOI: 10.2523/IPTC-18279-MS

[91] Stumm W. Reactivity at the mineral-water interface: Dissolution and inhibition. Colloids and Surfaces A: Physicochemical and Engineering Aspects. 1997;**120**(1-3):143-166. DOI: 10.1016/S0927-7757(96)03866-6

[92] Ng JH, Almubarak T, Nasr-El-Din HA. Low-carbon-steel corrosion at high temperatures by aminopolycarboxylic acids. SPE Production & Operations. 2017;**33**(01):131-144. DOI: 10.2118/188007-PA

[93] Al-Mutairi SH, Nasr-El-Din HA, Al-Driweesh S, Al-Muntasheri GA. Corrosion control during acid fracturing of deep gas wells: Lab studies and field cases. In: Paper presented at the SPE International Symposium on Oilfield Corrosion, Aberdeen, United Kingdom. May 2005. DOI: 10.2118/94639-MS

[94] Kalfayan L. Production enhancement with acid stimulation. Tulsa, Oklahoma, USA: Pennwell Books; 2008

[95] De Wolf C, Bouwman AB, Nasr-El-Din HA. Corrosion Resistance when using Chelating Agents in Carbon-Steel Environment. US Patent No. 20150005216 A1. 2015

[96] de Wolf CA, Rudolf E, Bouwman AF, Braun W, De P, Nasr-El-Din HA. Evaluation of environmentally friendly chelating agents for applications in the oil and gas industry. In: Paper presented at the SPE International Symposium and Exhibition on Formation Damage Control, Lafayette, Louisiana, USA. Feb 2014. DOI: 10.2118/168145-MS

[97] Nasr-El-Din HA, Corine, Bouwman A, Bang E, Naylor E, Alex A. A new, low corrosive fluid to stimulate wells with carbon steel tubular and internals. In: Paper presented at the SPE Saudi Arabia Section Technical Symposium and Exhibition, Al-Khobar, Saudi Arabia. Apr 2012. DOI: 10.2118/160849-MS

[98] Blesa MA, Borghi E, Maroto G, Regazzoni AE. Adsorption of EDTA and iron—EDTA complexes on magnetite and the mechanism of dissolution of magnetite by EDTA. Journal of Colloid and Interface Science. 1984;**98**(2):295-305. DOI: 10.1016/0021-9797(84)90155-3

[99] Ameri A, Nick HM, Ilangovan N, Pęksa A. A comparative study on the performance of acid systems for high temperature matrix stimulation. In: Paper presented at the Abu Dhabi International Petroleum Exhibition & Conference, Abu Dhabi, UAE. Nov 2016. DOI: 10.2118/183399-MS

[100] Martell AE, Motekaitis RJ, Fried AR, Wilson JS, Macmillan D.

Thermal decomposition of EDTA, NTA, and nitrilotrimethylenephosphonic acid in aqueous solution. Canadian Journal of Chemistry. 1975;**53**(22):3471-3476. DOI: 10.1139/v75-498

[101] Booy M, Swaddle TW. Chelating agents in high temperature aqueous chemistry. 2. The thermal decomposition of some transition metal complexes of nitrilotriacetate (NTA). Canadian Journal of Chemistry. 1977;**55**(10):1770-1776. DOI: 10.1139/v77-248

[102] Frenier W, Brady M, Al-Harthy S, Arangath R, Chan KS, Flamant N, et al. Hot oil and gas wells can be stimulated without acids. SPE Production & Facilities. 2004;**19**(04):189-199. DOI: 10.2118/86522-PA

[103] Sokhanvarian K. Thermal Stability of Various Chelates That Are Used in the Oil Field. College Station: Texas A&M; 2012

[104] Nowack B. Environmental chemistry of aminopolycarboxylate chelating agents. Environmental Science & Technology. 2002;**36**(19):4009-4016. DOI: 10.1021/es025683s

[105] Sillanpää MET, Agustiono Kurniawan T, Lo W. Degradation of chelating agents in aqueous solution using advanced oxidation process (AOP). Chemosphere. 2011;**83**(11):1443-1460. DOI: 10.1016/j.chemosphere.2011.01.007

[106] Sillanpää M. In: Ware GW, editor. Reviews of Environmental Contamination and Toxicology. New York: Springer Science; 1997. pp. 85-111

[107] Frenier WW, Wilson DS, Crump D, Jones LM. Use of highly Acid-Soluble chelating agents in well stimulation services. In: Paper presented at the SPE Annual Technical Conference and Exhibition, Dallas, Texas. Oct 2000. DOI: 10.2118/63242-MS

[108] Sýkora V, Pitter P, Bittnerová I, Lederer T. Biodegradability of ethylenediamine-based complexing agents. Water Research. 2001;**35**(8):2010-2016. DOI: 10.1016/S0043-1354(00)00455-3

[109] Pitter P, Chudoba J. Biodegradability of Organic Substances in the Aquatic Environment. 1st ed. Florida: CRC Press; 1990

[110] Sokhanvarian K, Nasr-El-Din HA, De wolf C. Thermal stability of oilfield aminopolycarboxylic acids/salts. SPE Production & Operations. 2016;**31**(01):12-21. DOI: 10.2118/157426-PA

Chapter 3

Schiff Bases and Their Metal Complexes in Solar Cells

Mirjana M. Radanović and Marijana S. Kostić

Abstract

Schiff bases represent a large group of organic compounds interesting for many different profiles of researchers due to their easy synthesis, versatile coordination behavior, and structural properties of their metal complexes, but primarily due to different application possibilities. Besides the promising biological activities, Schiff bases and their metal complexes often show high photoluminescence, thus making good candidates for use in optical materials. Among these, the use of Schiff base metal complexes in different types of solar cells stands out. With the aim to make this more attractive for more coordination chemists, in this chapter, we highlight the main findings from this field to establish a better understanding of the structure-properties linkage and enable the design of new materials with enhanced characteristics.

Keywords: coordination compounds, structure, schiff bases, photovoltaics, DSSC, perovskite solar cells, organic solar cells

1. Introduction

Countless efforts are being made to achieve better substitutes for fossil fuels and suppress air pollution and climate change [1]. One of the roads to a greener future is undoubtedly the utilization of solar light instead of fossil fuels and its conversion into electrical energy through solar cells.

Dye-sensitized solar cells (DSSC) represent the third generation of solar cells. During the 3 decades since their introduction, different classes of compounds were investigated as photosensitizers to achieve better efficiency, stability, and improved production procedures. Research has revealed that metal complexes play a crucial role in enhancing the stability and photovoltaic performance of DSSC. Compounds featuring double carbon-nitrogen bonds emerge as promising candidates for investigating optical properties [2]. The conjugation and delocalization of such bonds and their complexation to metal ions, thus forming so-called push-pull molecules, offer an added boost to their optical activity, making them attractive for applications in photovoltaic cells.

Initially, ruthenium complexes with various ligands were explored for this purpose. However, the current thrust is toward discovering materials that leverage more cost-effective and readily available metals, considering ruthenium's rarity and price. Research has shown that complexes of 3d and 4d metals with N-heterocyclic ligands have broad absorption spectra and excellent light-harvesting properties [1], making them promising materials for enhancing the sensitivity of DSSC.

In addition to the DSSC, the usage of Schiff base derivatives in organic solar cells and perovskite solar cells (PSC) is currently being explored. The latter, being the most attractive type of solar cells for researchers nowadays, opens a new path of combining the advantages of both types of materials.

Hence, this chapter will present the main findings in using Schiff base metal complexes for various purposes in different kinds of solar cells.

2. Schiff bases and their complexes in DSSC

One of the main advantages of DSSC was using dyes as the absorbing layer instead of the semiconductor. With this aim, different classes of compounds were investigated, such as natural sensitizers, polymers, derivatives of anthracene, and metal complexes, the latter being among the most promising ones [2, 3]. Organic dyes were more efficient due to unlocalized electrons, but inorganic dyes were shown to be more durable and stable at higher temperatures [4]. This made metal-organic derivatives more interesting for research since they had both advantages. Also, the complexation of Schiff bases to metal ions could enhance their optoelectronic properties [5] and make them more suitable for this application.

The research has shown that metal complexes used as sensitizers should contain anchoring ligands (usually carboxylic or phosphorous acid) that bind to the surface of the semiconductor and ancillary ligands with π-donor properties to enable a "push-pull" design [6–8].

One of the first complexes used as light absorbers was the ruthenium complex with bipyridine $[Ru(bpy)]^{2+}$. Its application resulted in low-efficiency solar cells due to their low stability [2]. Later studies were made to obtain acceptable power conversion efficiency (PCE), and some ruthenium polypyridyl complexes were proved to possess many advantages, such as wide absorption due to metal-to-ligand charge transfer, longer excitation lifetime, and long-term chemical stability. However, issues like poor absorption in near-IR, low yields, complicated purification procedures, the toxicity of the Ru-based materials, high-cost fabrication, and limited resources remained. This motivated the scientists to search for sensitizers that contain less toxic and easily available metal center or metal-free sensitizers. The latter was shown to have lower PCE, shorter excitation lifetime, and aggregation-induced self-quenching, similar to natural dyes [9].

Numerous complexes with different metals were synthesized and tested for this application, but our attention is drawn to the Schiff bases and their metal complexes. These compounds could overcome the mentioned drawbacks due to their suitable optoelectronic features, better stability, and environment-friendly synthetic routes.

Two complexes of copper(II) (**Figure 1**) and cobalt(III) with the NO bidentate Schiff base salicylidene-ethanolamine, used as sensitizers of DSSC [10], gave PCEs of 3 and 3.84%, respectively.

The majority of the examined compounds are mononuclear, often homoleptic complexes. Polymeric complexes are not thoroughly investigated, probably due to more complicated preparation procedures [11]. One of the newest results in this field is the synthesis and characterization of copper(II) and cadmium(II) complexes (**Figure 2**) containing one of these two salicylaldehyde Schiff base derivatives with a "push-pull" structure [8].

When used as sensitizers, cadmium(II) complexes were more efficient than copper(II) complexes with the same ligand, most probably due to the larger radius of the cadmium ion. The range of PCEs was from 4.77 to 8.59% [8].

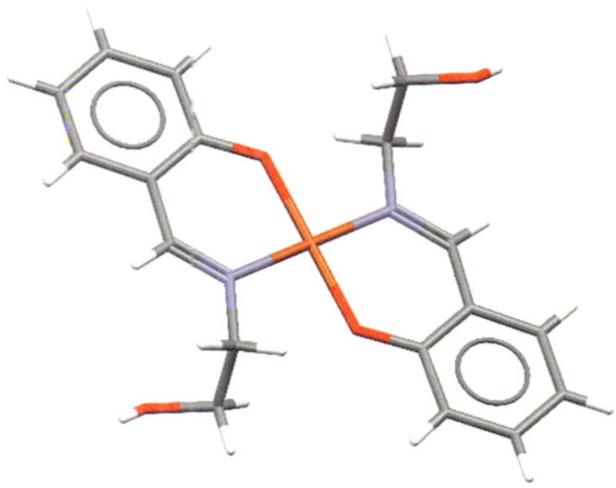

Figure 1.
Molecular structure of the copper(II) complex with Schiff base of salicylaldehyde and ethanolamine (refcode HESLCU10).

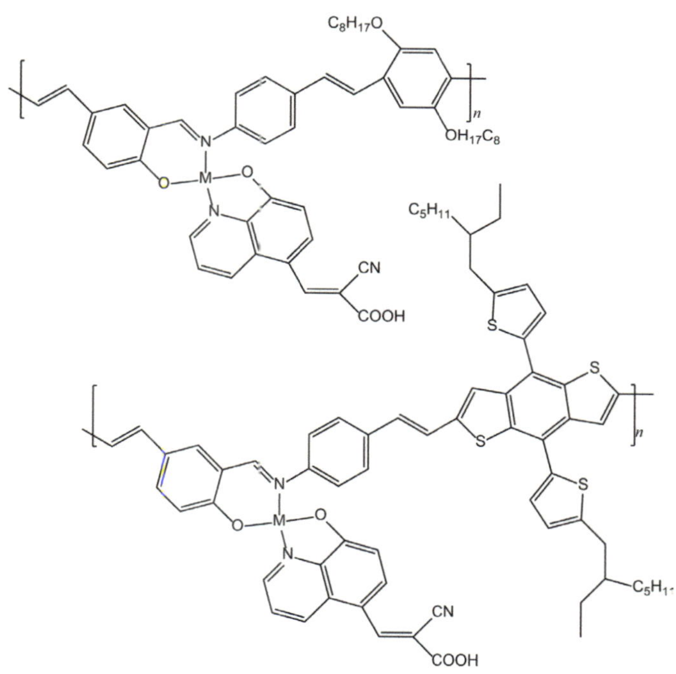

Figure 2.
Structural formulas of the complexes with salicylaldehyde derivative (M = Cu, Cd).

Other four complexes of d-metals were synthesized and characterized, and their efficiency as dyes was examined (**Figure 3**) [12]. Cobalt(II) and copper(II) complexes were less effective than zinc and cadmium ones, meaning the impact of central ion is one of the crucial factors for obtaining higher efficiency.

Copper complexes are among the most promising ones since they could be used not only as photosensitizers but also as redox shuttles (**Figure 4**) [13]. Namely, the commonly used triiodide/iodide redox couple, aside from good performance, could cause various problems due to its volatile nature, corrosive effect, and dark red color of the triiodide ion. These obstacles could be vanquished by using metal complexes with variable oxidation states.

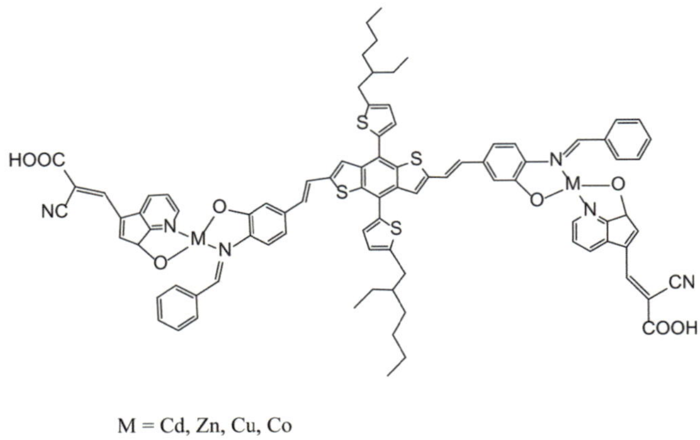

Figure 3.
The structural formula of complexes with thienylbenzo[1,2-b,4,5-b′]dithiophene and the 8-quinolinol derivative [12].

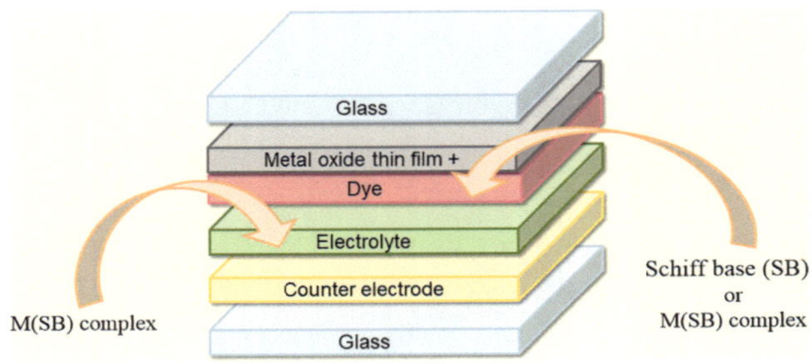

Figure 4.
Strategies to enhance DSSCs using Schiff bases or their metal complexes.

With this aim, cobalt complexes (**Figure 5**) were investigated, and some improvements were observed [14].

Still, because of limitations and potential health problems cobalt complexes could cause, copper(I/II) compounds of distorted tetragonal geometry were suggested as suitable. In addition to copper and cobalt complexes, iron, nickel, manganese, and vanadium complexes were also investigated as replacements for the triiodide/iodide redox couple [15].

In [16], an oxovanadium(IV/V) redox coupled with a well-known salen ligand (symmetric Schiff base of salicylaldehyde and ethylenediamine in molar ratio 2:1)

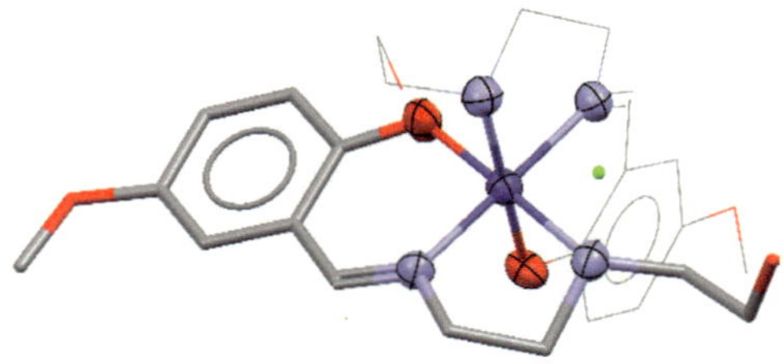

Figure 5.
Molecular structure of the bis(ligand) cobalt(III) complex (refcode YIWCAK). Hydrogen atoms are omitted for clarity; metal center and ligator atoms are represented as ellipsoids.

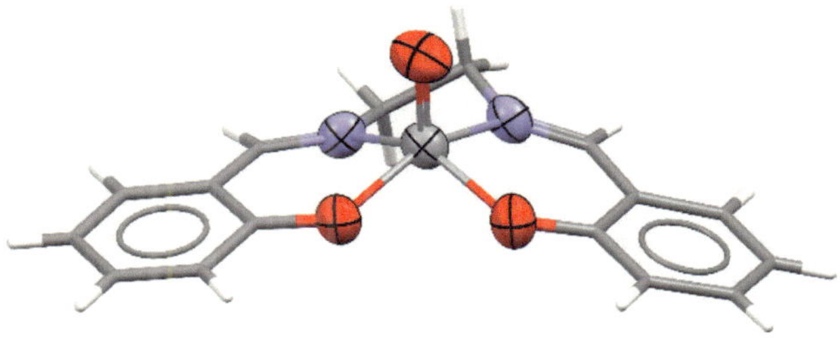

Figure 6.
Molecular structure of VO(salen) complex (refcode DOGSOH). Metal center and ligator atoms are represented as ellipsoids.

was used as a redox mediator in DSSC (**Figure 6**). The results showed the efficiency of back then satisfying more than 5%.

Until now, studies for optimal sensitizers have been much broader and more thorough than the search for alternatives to common redox shuttles. Still, this area represents a field in which much progress could be made in the following years.

3. Schiff bases and their metal complexes in perovskite solar cells

As said earlier, PSCs are currently the most attractive type of solar cells due to their tunable properties, high sensitivity, and high tolerance to structure defects, imperfections, and impurities. Besides, PSCs are simpler to fabricate and cost less than the most widely used and commercialized crystalline solar cells. The researchers' great interest in PSCs led to fast progress in the field—in 10 years, the PCE was enlarged almost seven times (from 3.8 to 25.8%) [17]!

Even though PSCs seem the best for now, they are far from ideal. One strategy for improvement consists of changing the hole-transport materials (HTM). Different compounds have been tested as HTMs, some purely organic—different small molecules, as well as polymers, some hybrid organic-inorganic and pure inorganic compounds such as different metal oxides, copper(I)-iodide, and copper(I)-thiocyanate. However, the PSCs with the best performances are produced using the spirobifluorene derivative Spiro-OMeTAD (**Figure 7**) as HTM.

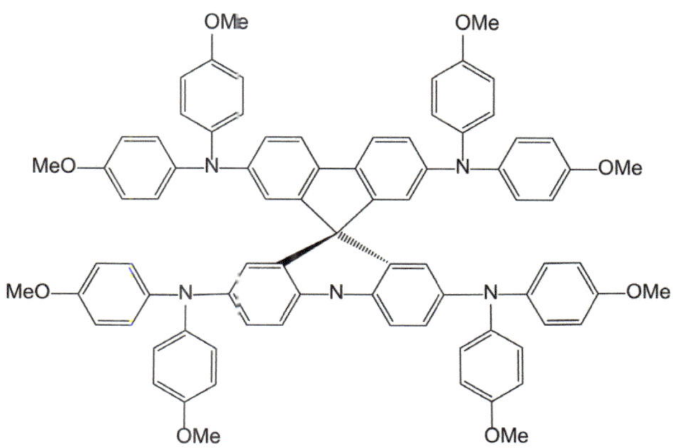

Figure 7.
The structural formula of the most effective HTM in PSCs.

The most commonly used HTMs involve complicated reaction routes with an inert atmosphere, high-cost catalysts, and a purification process. For this purpose, Schiff bases were suggested as a good replacement [18]. They are obtained in excellent yields in a simple one-pot condensation reaction, with water as the only by-product. This makes their synthesis easy, cost-effective, environment-friendly, and easy to scale up, while their purification process is straightforward. When applied as HTMs, the solar cell's performance is comparable to those with the usually used materials.

One of the drawbacks of Schiff bases is the hydrolysis azomethine bond is prone to. Nevertheless, this phenomenon is more noticeable in aliphatic derivatives, whereas in aromatic, highly conjugated ones, the imine group exhibits greater resistance to hydrolysis [19].

Some azomethines and diimines (**Figure 8**) were examined regarding acid-base discoloration and optoelectronic properties [20] and suggested as promising candidates for HTMs in PSCs but not competitive with Spiro-OMeTAD.

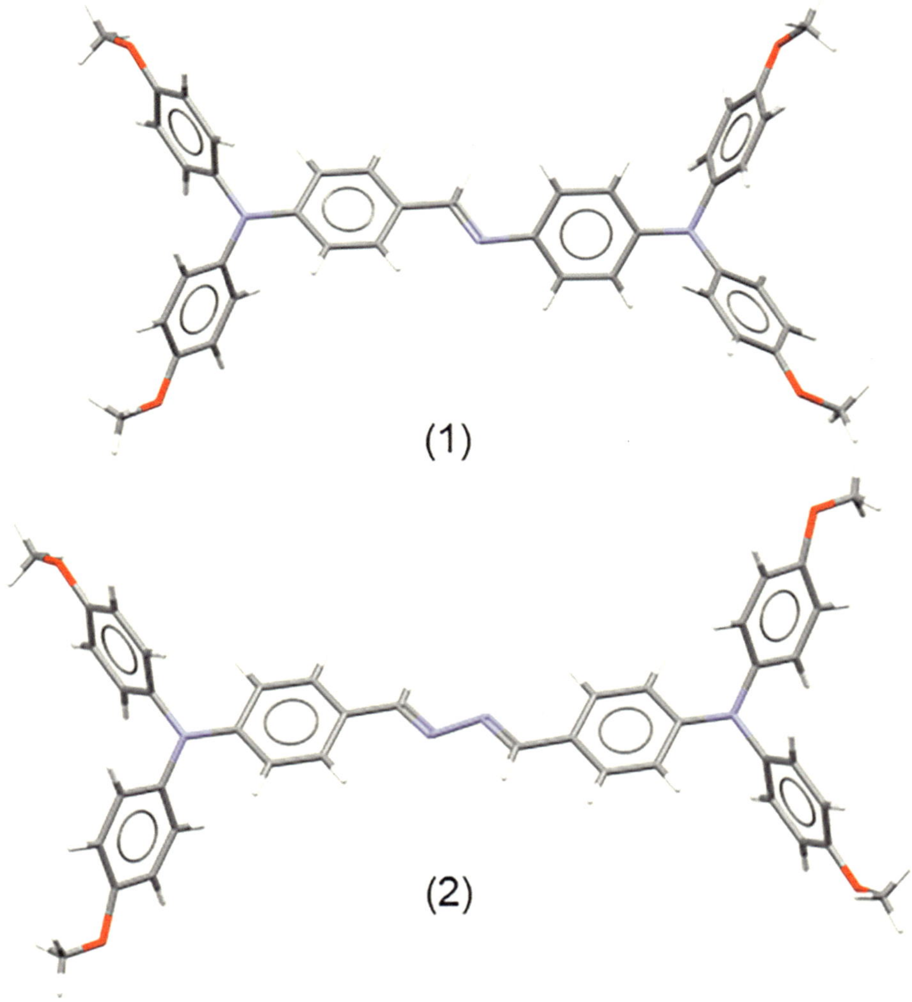

Figure 8.
Molecular structures of Schiff base derivative examined for use in PSCs (refcodes CUHXUA (1) and CUHXOU (2)).

Another series of Schiff bases as possible HTMs were synthesized, in which the core and direction of the azomethine bond were varied to examine their influence on the performance of PSCs [19]. The best ones showed to be those containing heterocyclic rings, most probably due to the stronger dipole and close molecular packing. Unlike Spiro-OMeTAD, all Schiff base derivatives used as HTMs were good against degradation due to moisture.

To overcome the disadvantages of state-of-the-art PSCs, some phenothiazines were also examined as potential HTMs (**Figure 9**). Namely, the existing method for HTM preparation typically relies on using predominantly chlorinated toxic solvents [21, 22]. These Schiff bases were synthesized in a procedure designed to minimize health and environmental hazards. Among these, AZO-II showed the best performance, with a PCE of 14%, excellent stability, and a low production price. While the results have not demonstrated a significant enhancement in efficiency, adopting a procedure nearly ten times cheaper and utilizing green solvents leads the authors to believe that this represents a promising approach for developing improved PSCs.

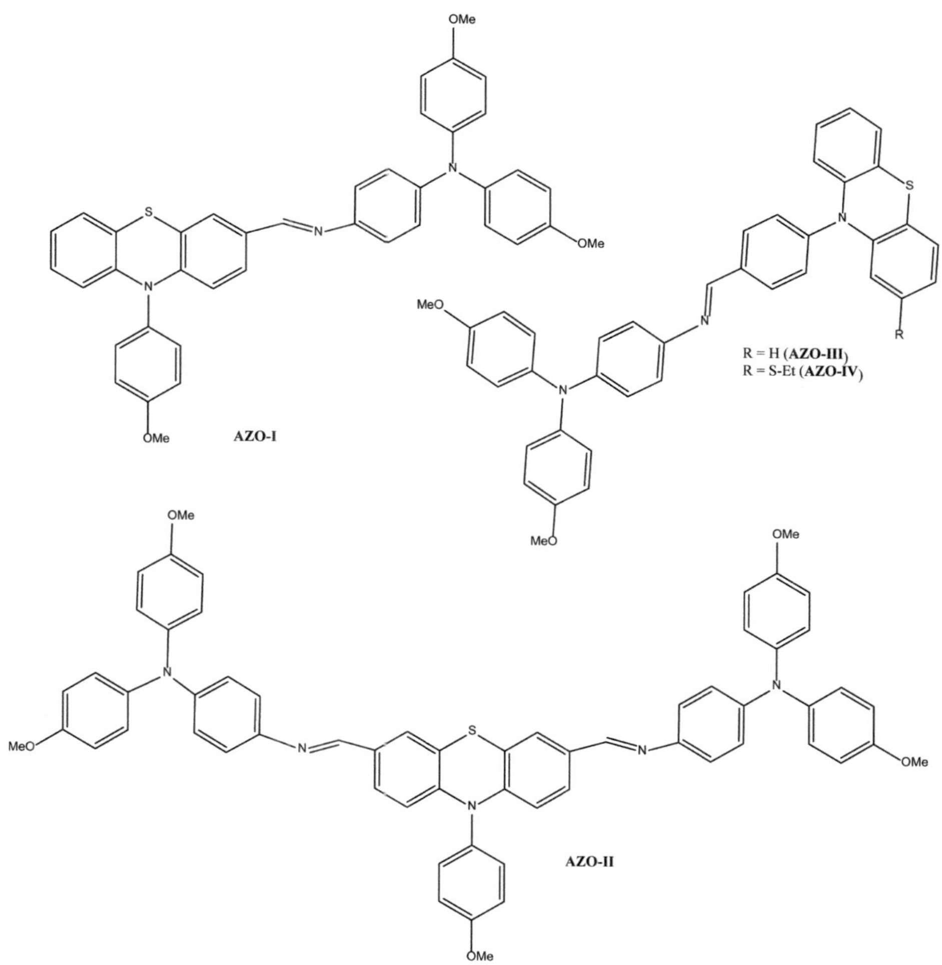

Figure 9.
Molecular structures of Schiff base derivative examined for use in PSCs.

A few years ago, interest in implementing Schiff base metal complexes to enhance the properties of perovskite solar cells arose. Namely, since the energy levels could be refined by changing the central metal ion, better alignment with the PSC absorber, lower surface roughness, and higher crystallinity could be achieved. All of this could lead to higher conversion efficiencies. Ones of the first to explore were complexes of platinum, palladium, and copper (**Figure 10**) [23]. These results open a new path to enhancing PSCs and solar cells in general.

Some earlier synthesized and characterized Co(II) complexes with the Schiff base of ninhydrin and glycin, obtained by template reaction (**Figure 11**), were evaluated as possible sensitizers for TiO_2 and ZnO layers [24]. The preliminary research has shown the enhanced efficiency of sensitized PSCs for a few percentiles.

One of the newest studies examined the use of Cu(II) complex with Schiff base derivative of 3,5-dichlorosalicylaldehyde and 3,4-diaminobenzoic acid as HTM [25] in all-inorganic PSCs, such as $CsBrBr_3$. Inorganic-organic hybrid PSCs (e.g., $CH_3NH_3PbX_3$ ($MAPbX_3$), $HC(NH_2)$ BbX_3($FAPbX_3$), and X = Cl, Br or I)) are more prone to moisture and oxygen invasion; thus, apart from reaching the highest efficiency, improved stability is necessary. This could be achieved by the enlargement of the inorganic portion of the hybrid and the use of Schiff base derivatives as HTMs. The outcome was the efficiencies comparable to those of carbon-based $CsBrBr_3$ cells, with the hint for future research to deal with using other transition metals and tuning the properties of materials by varying the Schiff base derivatives.

Generally, two approaches for utilizing Schiff bases and their metal complexes to improve PSCs are known so far: using these compounds as HTMs or as sensitizers for the electron transport layer, as summarized in **Figure 12**.

However, still, many issues remain to be addressed. One of them is the toxicity of the used PSCs, mostly because of the lead presence [17]. The global use of these solar cells could cause the contamination of the ecosystem with lead, which will eventually be present in the air, soil, and water in higher concentrations. Also, the higher stability of PSCs under high temperatures and humidity is of great interest. Considering the positive effect of Schiff bases and their metal complexes on improving the material's resilience to corrosion [26–28] and the fact that the application of metal complexes in PSCs is not thoroughly examined, this path could be one of the solutions for the mentioned issues.

Besides, some novel paths in designing the structure of the complexes might be attractive based on the results of the assessment of optoelectronic properties of some Schiff bases of aminoguanidine [29–31].

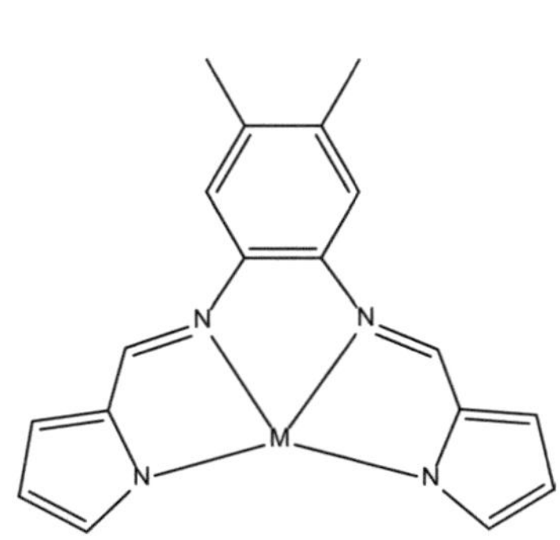

Figure 10.
Structural formula of the complexes (M = Pt, Pd, Cu).

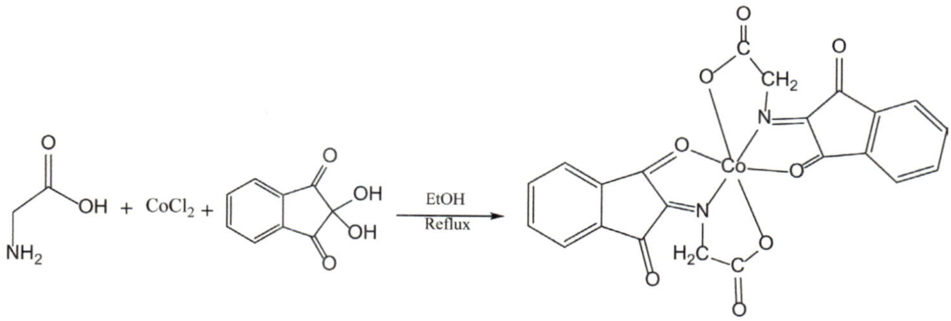

Figure 11.
Reaction route for the synthesis of Co(II) complex with Schiff base of ninhydrin and glycin.

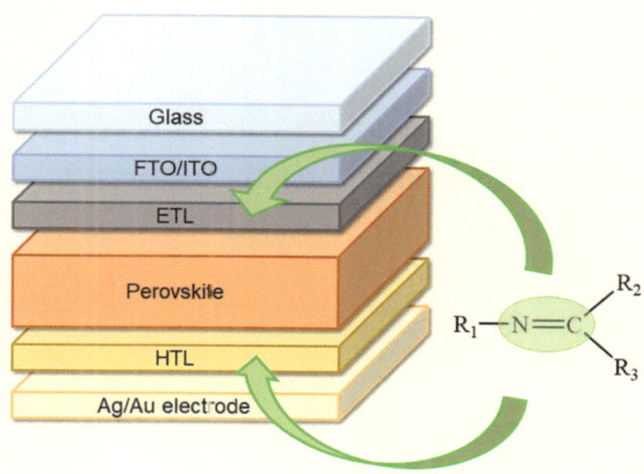

Figure 12.
Examined possibilities of enhancement of PSCs by using Schiff bases and their metal complexes. FTO = fluoride doped tin oxide; ITO = indium tin oxide; ETL = electron transport layer, HTL = hole transport layer.

Also, it should be noted that detailed structural characterization of these materials would lead to better insight into structure-properties linkage, which could enable the design of environment-friendly stable solar products with optimal efficacy and lifetime, as well as manufacturing costs.

4. Conclusions

Utilizing Schiff bases and their metal complexes in solar cells offers promising potential for advancing photovoltaic technology. Exploring these materials in dye-sensitized solar cells and perovskite solar cells revealed they provide notable advantages like adjustable electronic properties, high light absorption, and easy manufacturing processes. Additionally, exploring novel Schiff base ligands and metal complexes holds promise for enhancing device efficiency, stability, and cost-effectiveness while considering major environmental challenges. Nevertheless, more detailed studies are needed to understand and optimize their performance in solar cell applications fully. In both dye-sensitized solar cells and perovskite solar cells, the interaction between the Schiff base ligands and metal ions impacts the device's efficiency and stability. Thus, the design of novel Schiff bases and their complexes should be highly interesting. Future research should focus on comprehensive structural characterization by X-ray crystallography and the comparative analysis of the structures and performances of numerous compounds to gain better insight into the structure-property relationships governing their performance.

Acknowledgements

The authors gratefully acknowledge the financial support of the Ministry of Science, Technological Development and Innovation of the Republic of Serbia (Grants No. 451-03-66/2024-2103/200125 & 451-03-65/2024-2103/200125).

Notes/thanks/other declarations

Molecular structures presented throughout the text are visualized using Mercury software.

Author details

Mirjana M. Radanović* and Marijana S. Kostić
Faculty of Sciences, University of Novi Sad, Novi Sad, Serbia

*Address all correspondence to: mirjana.lalovic@dh.uns.ac.rs

References

[1] Colombo A, Dragonetti C, Roberto D, Fagnani F. Copper complexes as alternative redox mediators in dye-sensitized solar cells. Molecules. 2021;**26**:194. DOI: 10.3390/molecules26010194

[2] Polo AS, Itokazu MK, Murakami Iha NY. Metal complex sensitizers in dye-sensitized solar cells. Coordination Chemistry Reviews. 2004;**248**:1343-1361. DOI: 10.1016/j.ccr.2004.04.013

[3] Kalyanasundaram K, Grätzel M. Applications of functionalized transition metal complexes in photonic and optoelectronic devices. Coordination Chemistry Reviews. 1998;**177**:347-414. DOI: 10.1016/S0010-8545(98)00189-1

[4] Arjmand F, Rashidi Ranjbar Z, Fatemi EGH. Effect of dye complex structure on performance in DSSCs; An Experimental and Theoretical Study. Heliyon. 2022;**8**:e11692. DOI: 10.1016/j.heliyon.2022.e11692

[5] Chouk R, Aguir C, Haouanoh D, Bergaoui M, Tala-Ighil R, Stathatos E, et al. A first-principles computational and experimental investigation on Schiff base cobalt complex towards designing solar cells. Journal of Molecular Structure. 2019;**1196**:676-684. DOI: 10.1016/J.MOLSTRUC.2019.07.026

[6] Cantón-Díaz AM, Muñoz-Flores BM, Moggio I, Arias E, De León A, García-López MC, et al. One-pot microwave-assisted synthesis of organotin Schiff bases: An optical and electrochemical study towards their effects in organic solar cells. New Journal of Chemistry. 2018;**42**:14586-14596. DOI: 10.1039/C8NJ02998A

[7] Lüthi E, Cortés PAF, Prescimone A, Constable EC, Housecroft CE. Schiff base ancillary ligands in bis(diimine) copper(I) dye-sensitized solar cells. International Journal of Molecular Sciences. 2020;**21**:1735. DOI: 10.3390/IJMS21051735

[8] Tian Y, Wang K, Zhang H, Wu X, Zhong C. Novel polymeric metal complexes of salicylaldehyde Schiff base derivative being used for dye sensitizer. Tetrahedron. 2022;**113**:132756. DOI: 10.1016/J.TET.2022.132756

[9] Mahadevi P, Sumathi S. Mini review on the performance of Schiff base and their metal complexes as photosensitizers in dye-sensitized solar cells. Synthetic Communications. 2020;**50**:2237-2249. DOI: 10.1080/00397911.2020.1748200

[10] Gautam C, Srivastava D, Kociok-Köhn G, Gosavi SW, Sharma VK, Chauhan R, et al. Copper(II) and cobalt(III) Schiff base complexes with hydroxy anchors as sensitizers in dye-sensitized solar cells (DSSCs). RSC Advances. 2023;**13**:9046-9054. DOI: 10.1039/D3RA00344B

[11] Okubo T, Tanaka N, Anma H, Kim KH, Maekawa M, Kuroda-Sowa T. Dye-sensitized solar cells with new one-dimensional halide-bridged Cu(I)–Ni(II) heterometal coordination polymers containing hexamethylene dithiocarbamate ligand. Polymers. 2012;**4**:1613-1626. DOI: 10.3390/POLYM4031613

[12] Zhang H, Wu X, Tang S, Wang K, Tian Y, Zhong C. Novel metal complexes for D-(A-π-a)2 motif dye sensitizer: Synthesis and photovoltaic application. Applied Organometallic Chemistry. 2021;**35**:e6220. DOI: 10.1002/AOC.6220

[13] Fagnani F, Colombo A, Dragonetti C, Roberto D. Recent investigations on the

use of copper complexes as molecular materials for dye-sensitized solar cells. Molecules. 2024;**29**:6. DOI: 10.3390/MOLECULES29010006

[14] Nasr-Esfahani M, Zendehdel M, Yaghoobi Nia N, Jafari B, Khosravi BM. Fabrication and characterization of a new dye sensitized solar cell with a new Schiff base cobalt complex as a redox mediator. RSC Advances. 2014;**4**:15961-15967. DOI: 10.1039/C3RA46531D

[15] Muñoz-García AB, Benesperi I, Boschloo G, Concepcion JJ, Delcamp JH, Gibson EA, et al. Dye-sensitized solar cells strike back. Chemical Society Reviews. 2021;**50**:12450-12550. DOI: 10.1039/D0CS01336F

[16] Oyaizu K, Hayo N, Sasada Y, Kato F, Nishide H. Enhanced bimolecular exchange reaction through programmed coordination of a five-coordinate oxovanadium complex for efficient redox mediation in dye-sensitized solar cells. Dalton Transactions. 2013;**42**:16090-16095. DOI: 10.1039/C3DT51698A

[17] Diouf B, Muley A, Pode R. Issues, challenges, and future perspectives of perovskites for energy conversion applications. Energies. 2023;**16**:6498. DOI: 10.3390/EN16186498

[18] Petrus ML, Bein T, Dingemans TJ, Docampo P. A low cost azomethine-based hole transporting material for perovskite photovoltaics. Journal of Materials Chemistry A. 2015;**3**:12159-12162. DOI: 10.1039/C5TA03046C

[19] Petrus ML, Music A, Closs AC, Bijleveld JC, Sirtl MT, Hu Y, et al. Design rules for the preparation of low-cost hole transporting materials for perovskite solar cells with moisture barrier properties. Journal of Materials Chemistry A. 2017;**5**:25200-25210. DOI: 10.1039/C7TA06452G

[20] Ma BB, Zhang H, Wang Y, Peng YX, Huang W, Wang MK, et al. Visualized acid–base discoloration and optoelectronic investigations of azines and azomethines having double 4-[N,N-di(4-methoxyphenyl)amino]phenyl terminals. Journal of Materials Chemistry C. 2015;**3**:7748-7755. DOI: 10.1039/C5TC00909J

[21] Salunke J, Guo X, Lin Z, Vale JR, Candeias NR, Nyman M, et al. Phenothiazine-based hole-transporting materials toward eco-friendly perovskite solar cells. ACS Applied Energy Materials. 2019;**2**:3021-3027. DOI: 10.1021/ACSAEM.9B00408

[22] Salunke J, Guo X, Liu M, Lin Z, Candeias NR, Priimagi A, et al. N-substituted phenothiazines as environmentally friendly hole-transporting materials for low-cost and highly stable halide perovskite solar cells. ACS Omega. 2020;**5**:23334-23342. DOI: 10.1021/acsomega.0c03184

[23] Wang Y, Ye W, Yang X, Rezaee E, Shan H, Yang S, et al. Hole transport layers based on metal Schiff base complexes in perovskite solar cells. Synthetic Metals. 2020;**259**:116248. DOI: 10.1016/J.SYNTHMET.2019.116248

[24] Chouk R, Haouanoh D, Bergaoui M, Aguir C, Tala-Ighil R, Khalfaoui M. Efficient planar perovskite solar cells using Schiff base complex as sensitizer for TiO2 and ZnO layers. In: 6th IEEE International Energy Conference (ENERGYCon 2020); 28 September 2020; Gammarth, Tunisia. New York, USA: IEEE; 2020. pp. 7-10

[25] Zheng L, Jackson E, Warren N, Owor R, Li Z. Perovskite CsPbBr3 solar cells with novel hole-transporting layer of metal complexes. Georgia Journal of Science. 2023;**81**:15

[26] Verma C, Quraishi MA. Recent progresses in Schiff bases as aqueous phase corrosion inhibitors: Design and applications. Coordination Chemistry Reviews. 2021;**446**:214105. DOI: 10.1016/J.CCR.2021.214105

[27] Afshari F, Ghomi ER, Dinari M, Ramakrishna S. Recent advances on the corrosion inhibition behavior of Schiff base compounds on mild steel in acidic media. ChemistrySelect. 2023;**8**:e202203231. DOI: 10.1002/slct.202203231

[28] Boulechfar C, Ferkous H, Delimi A, Berredjem M, Kahlouche A, Madaci A, et al. Corrosion inhibition of Schiff base and their metal complexes with [Mn (II), Co (II) and Zn (II)]: Experimental and quantum chemical studies. Journal of Molecular Liquids. 2023;**378**:121637. DOI: 10.1016/J.MOLLIQ.2023.121637

[29] Jelić MG, Georgiadou DG, Radanović MM, Romčević N, Giannakopoulos KP, Leovac VM, et al. Efficient electron injecting layer for PLEDs based on (PLAGH)2[ZnCl4]. Optical and Quantum Electronics. 2016;**48**:1-10. DOI: 10.1007/s11082-016-0547-5

[30] Radanović MM, Vojinović-Ješić LS, Jelić MG, Sakellis E, Barta Holló B, Leovac VM, et al. Synthesis, structures, and photoluminescence of two novel zinc(II) compounds containing 2-acetylpyridine-aminoguanidine. Inorganics. 2022;**10**:147. DOI: 10.3390/INORGANICS10100147

[31] Babic B, Romcevic M, Gilic M, Trajic J, Radanović MM, Vojinović-Ješić LS, et al. Phonon assisted charge transfer in complexes of Zn(II) and Cd(II) with 2-acetylpyridine-aminoguanidine. Optical Materials. 2023;**136**:113445. DOI: 10.1016/J.OPTMAT.2023.113445

Section 2

Innovations in Chromatography, Microfluidics and Laboratory Diagnostics

Chapter 4

Advanced Chromatography Analytical Methods for the Isolation and Identification of Natural Drug Molecules

Paranthaman Subash, K.K. Senthil Kumar, Kareti Srinivasa Rao and Sulekha Khute

Abstract

The creation of stability-indicating analytical methods, phytoconstituent identification, and validation for content and impurity estimation in pharmaceutical drug products and drug substances using high-performance liquid chromatography (HPLC), ultra-performance liquid chromatography (UPLC), liquid chromatography-mass spectrometry (LC-MS), gas chromatography-mass spectrometry (GC-MS), and supercritical fluid chromatography (SFC) employing contemporary analytical techniques are the primary topics of this chapter. The intricacy and diversity of herbal components provide the most analytical hurdle. Effective analytical methods are thus crucial for the separation and qualitative and quantitative analysis of multi-component systems. Recent advances in selectivity, sensitivity, and speed of analysis have made a number of newly developed analytical approaches a major support for complex component analysis. This chapter provides an overview of the application of advanced chromatography for the separation and identification of natural components in herbal medicines. By carefully choosing and refining both fixed and mobile phases, the scope of applications was increased.

Keywords: HPLC, UPLC, LC-MS, GC-MS, SFC

1. Introduction

1.1 Natural products in drug discovery

Natural drug molecules are bioactive compounds that are derived from a variety of natural sources, including plants, fungi, and microorganisms [1]. These compounds have been harnessed for their therapeutic properties for centuries. In traditional medicine, natural compounds have been used to treat a wide range of ailments [2]. Many cultures have used plants for medicinal purposes. The willow tree, for instance, contains salicin, a precursor to aspirin. Penicillin, one of the most famous antibiotics, was derived from the *Penicillium* fungus. Streptomycin, an antibiotic used to treat tuberculosis, was isolated from the soil bacterium *Streptomyces griseus* [3]. In modern medicine, natural drug molecules continue to be a significant source of new pharmaceuticals. Researchers often look to nature for inspiration when developing new drugs. Here are some notable isolated secondary metabolite examples and its sources listed in **Table 1**.

1.2 High-performance liquid chromatography (HPLC)

HPLC has emerged as a cornerstone in the analytical chemistry toolbox, particularly in the development, isolation, and identification of natural drug molecules [14]. Its unparalleled sensitivity, high resolution, and versatility have made it a preferred method in pharmaceutical research, environmental monitoring, and food safety analysis. This essay delves into the significance of advanced chromatography analytical methods, particularly focusing on their applications in natural drug development. The evolution of chromatography has led to the advancement of HPLC, enabling it to separate complex mixtures with remarkable efficiency. In the context of natural drug molecules, which often exhibit structural complexity and a diverse range of chemical properties, HPLC serves as a critical method for isolating active pharmaceutical ingredients (APIs) from plant and other biological sources. By employing different mobile phases and stationary phases, analysts can optimize conditions to achieve optimal separation of compounds, including those that may be present in trace amounts [15].

For instance, the application of reversed-phase HPLC (RP-HPLC) techniques has greatly enhanced the isolation process of hydrophobic compounds, often found in natural products. By utilizing high-pressure pumps and sophisticated detector systems, analysts can fine-tune the gradient elution process, allowing for better resolution of compounds that may co-elute in traditional methodologies [16]. This precision is crucial for researchers aiming to isolate specific bioactive compounds necessary for further pharmacological evaluation. Following isolation, the identification of drug molecules requires equally sophisticated methods. Advanced HPLC

Chromatography Technique	Secondary Metabolite	Chemical Class	Example Source	Reference
HPLC	Quercetin	Flavonoid	Apples and onions	[4]
	Caffeine	Alkaloid	Coffee beans	[5]
UPLC	Gallic Acid	Phenolic Acid	Green tea	[6]
	Limonene	Terpene	Citrus fruits	[7]
LCMS	Anthocyanins	Glycoside	Berries (e.g., blueberries)	[8]
	Ginsenosides	Saponin	Ginseng	[9]
GCMS	Eucalyptol	Monoterpene	Eucalyptus oil	[10]
	Linoleic Acid	Fatty Acid	Sunflower oil	[11]
SFC	Carotenoids	Terpenoid	Carrots and tomatoes	[12]
	Phytosterols	Steroid	Plant oils	[13]

Table 1.
Isolated secondary metabolite examples and its sources.

can be coupled with mass spectrometry (MS), a combination known as HPLC-MS, which provides structural elucidation and quantification of compounds. The ability to analyze the mass-to-charge ratio of ionized particles significantly enhances structural identification, allowing analysts to confirm the identity of isolated compounds rapidly. Moreover, in the realm of natural product research, the use of two-dimensional chromatography (2D-HPLC) has gained traction. This method, which utilizes two different separation techniques in tandem, enables an even more sophisticated analysis of complex mixtures. Such approaches facilitate the separation of metabolites from crude extracts, offering insights into the pharmacokinetics and pharmacodynamics of components, which are vital for drug formulation [17].

Aside from HPLC's technical capabilities, its application is further extended through method development strategies tailored for specific natural drug molecules. This includes the optimization of extraction processes, where solvents and extraction times are meticulously selected to maximize yield while preserving compound integrity. The application of supervised machine learning algorithms for designing and optimizing HPLC methods represents a forward-thinking approach, guiding analysts through the process of achieving reproducible and reliable results in compound isolation and identification [18]. The implications of these advanced analytical methods extend far beyond mere academic interest. As the demand for natural drugs grows within the pharmaceutical industry, regulatory bodies also require stringent testing and verification of safety and efficacy. The advancements in chromatography analytical methods, particularly HPLC, have revolutionized the landscape of drug development, isolation, and identification. With its unmatched efficiency and versatility, HPLC supports researchers in isolating and identifying natural drug molecules, paving the way for future pharmaceutical innovations. As technology continues to advance, the integration of HPLC with other analytical techniques will further enhance its capabilities, solidifying its position as an indispensable tool in the analytical chemistry field [14].

1.3 Method development of HPLC

1.3.1 System suitability tests

Before embarking on method development, researchers conduct system suitability tests to ensure that the HPLC system is functioning within acceptable parameters. These tests are crucial as they assess key performance indicators such as resolution, peak symmetry, and retention time variability. By evaluating these parameters, researchers establish a baseline that confirms the reliability and reproducibility of the analytical method. Essentially, system suitability tests act as a quality control measure, ensuring that the HPLC system is capable of producing consistent and accurate results, which is foundational for any subsequent analytical work [19].

1.3.2 Column selection

The choice of the chromatographic column is indeed pivotal in achieving the desired separation in any chromatographic analysis. Different columns possess unique stationary phases and dimensions, which significantly impact selectivity and efficiency. For instance, columns with polar stationary phases are ideal for separating polar compounds, while non-polar phases are better suited for non-polar analytes. The dimensions of the column, such as length and diameter, also play a crucial role; longer columns generally provide better resolution but at the cost of increased analysis time. Researchers must carefully consider these factors, along with the nature of the analytes and the objectives of the analysis, to select an appropriate column. This meticulous selection process ensures optimized separation, accurate results, and efficient use of resources [20].

1.3.3 Mobile phase composition

The composition of the mobile phase plays a critical role in determining the separation efficiency in HPLC. Researchers must carefully optimize the ratio and type of solvents used in the mobile phase, taking into account key factors such as polarity, solubility, and compatibility with the stationary phase. The choice of mobile phase directly impacts the retention time and resolution of the analytes, which are crucial for achieving accurate and precise results. By fine-tuning the mobile phase composition, scientists can enhance the selectivity and sensitivity of the separation process, ultimately leading to more reliable and reproducible analyses [21].

1.3.4 Flow rate optimization

The flow rate of the mobile phase through the column is a critical parameter affecting the speed and efficiency of the separation in chromatographic processes. Researchers strive to optimize this flow rate to balance the need for rapid analysis with the requirement for optimal resolution. Inadequate flow rates can lead to poor separation of the analytes, making it difficult to distinguish between different compounds. On the other hand, excessive flow rates may compromise the resolution by not allowing sufficient interaction time between the mobile phase and the stationary phase, thus reducing the effectiveness of the separation. Achieving the ideal flow rate is crucial for ensuring accurate and efficient analysis, and it often requires a careful balance and fine-tuning based on the specific characteristics of the compounds being analyzed [22].

1.3.5 Temperature control

Temperature control in HPLC systems, particularly within the column, is pivotal for achieving optimal separation of analytes. Maintaining a consistent temperature ensures reproducibility and stability of the chromatographic method. Elevated temperatures can accelerate the separation process, potentially reducing run times and improving peak shapes. However, this may also compromise the resolution of closely eluting compounds, as higher temperatures can decrease the viscosity of the mobile phase and alter the interactions between analytes and the stationary phase. Therefore, careful optimization of the temperature settings is essential to balance the speed of separation with the resolution and efficiency of the method, ensuring accurate and reliable analytical results [23].

1.3.6 Detector wavelength

Selecting the appropriate wavelength for detection is crucial, especially when utilizing UV-visible detectors in analytical chemistry. The detector wavelength must align with the absorption maxima of the analytes to ensure optimal sensitivity and accuracy. This alignment enables the detector to effectively measure the concentration of substances, as it corresponds to the wavelength where the analytes absorb the lightest. By carefully choosing the correct wavelength, one can achieve higher sensitivity, leading to more precise and reliable quantitative results. This parameter is indispensable in various applications, from pharmaceutical analysis to environmental monitoring, where accuracy and reliability are paramount [24].

1.3.7 Injection volume

In HPLC, the volume of the sample injected plays a crucial role in determining the sensitivity and peak shape of the chromatographic analysis. Researchers meticulously optimize the injection volume to achieve a delicate balance between maximizing signal intensity and avoiding column overloading. Overloading the column can lead to undesirable effects such as peak broadening, which can compromise the resolution and accuracy of the results. By carefully calibrating the injection volume, scientists ensure that the peaks remain sharp and well-resolved, thereby enhancing the reliability and efficiency of the HPLC method for quantitative and qualitative analysis [25].

1.3.8 pH adjustment

For ionizable compounds, adjusting the pH of the mobile phase is a crucial factor in optimizing chromatographic separation. The pH level directly influences the ionization states of analytes, which in turn affects their interactions with the stationary phase. By carefully controlling the pH, one can enhance the selectivity and resolution of the separation process. For instance, if the pH is adjusted to a level where the analyte is fully ionized, it may exhibit stronger or weaker interactions with the stationary phase depending on its charge and the nature of the stationary phase. Conversely, at a pH where the analyte is neutral, its retention time and separation behavior will differ. Therefore, pH optimization is a powerful tool in chromatographic techniques, enabling precise control over the separation of complex mixtures [26].

2. Ultra-performance liquid chromatography (UPLC)

In recent years, the field of pharmacognosy has witnessed an intensified focus on the development, isolation, and identification of natural drug molecules. Advanced chromatography analytical methods have played a pivotal role in these efforts, offering enhanced sensitivity, resolution, and specificity. Ultra-performance liquid chromatography (UPLC) stands out as a particularly effective technique, significantly advancing the analysis of complex mixtures of natural products [27]. UPLC, a refined form of traditional high-performance liquid chromatography (HPLC), utilizes smaller particle sizes, higher pressure, and advanced detectors, thereby improving the separation efficiency and speed of analyses. The key innovations in UPLC include the use of sub-2-micron column packing, which allows for a higher surface area and provides better interaction with the analytes. This enhancement leads to sharper peaks in chromatograms and faster analysis times, which is crucial when dealing with the often complex and varied compositions found in natural sources [7].

Isolation of natural drug molecules typically involves extracting compounds from biological sources, such as plants, fungi, and marine organisms. Advanced chromatography methods facilitate the extraction process, allowing for swift separation of compounds based on their chemical properties. For instance, during the initial stages of extraction, a sequential approach can be employed using different solvents in conjunction with UPLC to identify the optimal solvent that maximizes yield and enhances the quality of the isolated compounds. This strategy underscores the importance of method development tailored for specific natural products, ensuring that the most efficacious compounds are identified and isolated for further research [28]. The identification of natural drug molecules requires not only effective isolation but also precise characterization. UPLC, coupled with mass spectrometry (MS), has transformed the identification process. This combination provides detailed insights into the molecular weight and structure of the isolated compounds. UPLC-MS allows researchers to accurately determine the identity of complex mixtures through the analysis of their fragmentation patterns and retention times. Moreover, the rapid identification capabilities of UPLC mitigate the risks of degradation or transformation of sensitive natural products during analysis, ensuring that the molecular integrity of the compounds is maintained [29].

In addition to UPLC, other advanced chromatographic techniques, such as preparative chromatography and two-dimensional chromatography, have been employed to further improve the separation and quantification of natural drug molecules. Preparative chromatography enables the large-scale purification of specific compounds, which can then be subjected to biological assays for pharmacological evaluation. Two-dimensional chromatography offers enhanced resolution, allowing for the separation of closely related compounds that may have similar chemical properties. Utilizing these techniques in conjunction with UPLC provides a comprehensive toolkit for researchers looking to explore the vast potential of natural products as therapeutic agents [30]. The integration of UPLC and advanced chromatography analytical methods into the workflow of natural product research facilitates a more systematic approach to drug discovery. Researchers are now able to streamline the identification and characterization processes, reducing the time and resources required to isolate active compounds. This efficiency is particularly critical given the current emphasis on finding new therapeutic agents from natural sources to address the growing concern of antibiotic resistance and the need for novel treatments [31].

The development, isolation, and identification of natural drug molecules are undergoing a transformation driven by advanced chromatography analytical methods, particularly UPLC. The benefits of enhanced sensitivity, speed, and specificity make UPLC an invaluable tool in the pharmacognosy field. As the demand for new therapeutics continues to rise, embracing these advanced techniques will not only enable the efficient exploration of natural products but also pave the way for innovative drug development strategies that have the potential to revolutionize healthcare.

2.1 Method development of UPLC

2.1.1 Mobile phase composition

The composition of the mobile phase is critical for the separation process. It typically consists of a mixture of solvents, such as water and acetonitrile or methanol. The pH and ionic strength of the mobile phase can significantly impact the retention time and resolution of analytes. Buffer solutions are often used to maintain the pH within a specific range to enhance the consistency and reproducibility of the results [32].

2.1.2 Flow rate

The flow rate of the mobile phase through the column affects the analysis time and resolution. Higher flow rates can reduce the analysis time but may compromise the resolution and sensitivity. Conversely, lower flow rates can enhance resolution but increase the analysis time. Finding the optimal flow rate is a balancing act that requires careful consideration of the specific analytical needs [33].

2.1.3 Temperature

Column temperature can influence the viscosity of the mobile phase and the interaction between the analytes and the stationary phase. Higher temperatures can improve the efficiency and speed of the separation but may also affect the stability of temperature-sensitive compounds. Consistent temperature control is essential for reproducibility [34].

2.1.4 Column type

Selecting the appropriate column type is fundamental for achieving good separation. Columns vary in terms of their stationary phase, particle size, and dimensions. The choice of column should be based on the nature of the analytes and the desired resolution. Columns with smaller particle sizes generally offer higher resolution but may require higher pressures [35].

2.1.5 Detector selection

The choice of detector depends on the properties of the analytes and the required sensitivity. Common detectors used in UPLC include UV-Vis, fluorescence, and mass spectrometers. Each detector has its advantages and limitations, and the selection should align with the specific analytical goals [36].

2.1.6 Method validation

Method validation ensures that the UPLC method is accurate, precise, and reliable. Key parameters to validate include specificity, linearity, accuracy, precision, limit of detection (LOD), limit of quantification (LOQ), and robustness. Validation provides confidence that the method will perform consistently in routine analysis [37].

2.1.7 Software tools

Several software tools can aid in UPLC method development and optimization. Design of experiments (DoE) allows for systematic experimentation to determine the effect of multiple factors on the outcome. Quality by design (QbD) is a holistic approach that emphasizes understanding the process and controlling variability to ensure quality results [38].

3. Gas chromatography-Mass spectrometry (GC-MS)

Advanced chromatography analytical methods have become instrumental in the development, isolation, and identification of natural drug molecules. These methodologies not only enhance the efficiency of the separation processes but also significantly improve the accuracy of analyte quantification, which is critical in pharmaceutical development [39]. GC-MS is a prominent technique that exemplifies the synchronization of chromatography with mass spectrometry. In the context of natural drug molecules, GC-MS leverages the volatility of compounds, allowing for their effective separation owing to differences in their boiling points and affinities for the stationary and mobile phases. This method is particularly valuable for volatile organic compounds, often found in essential oils and plant extracts, enabling their precise identification and quantification [40].

In the development phase of natural drug molecules, GC-MS serves as a robust tool for screening crude extracts derived from plants or other natural sources. This capability is crucial, as it aids in identifying bioactive compounds that could be potential candidates for drug development. With the incorporation of advanced sample preparation techniques such as solid-phase microextraction (SPME), the efficiency is significantly improved, thus fine-tuning the sensitivity and specificity of the analysis [41]. The isolation of active constituents is another critical step in natural product chemistry. GC-MS facilitates this through its dual-functionality; not only does it provide qualitative data, but it also yields quantitative results that help in determining the concentration of specific compounds within a mixture. This is vital when isolating compounds that may exist in minimal concentrations, allowing researchers to ascertain their potential efficacy in therapeutic applications [42].

Moreover, the identification of compounds in complex matrices is made feasible by employing advanced data analysis techniques associated with GCMS. Software and databases that support spectral matching help researchers confidently distinguish between closely related natural products, therefore ensuring high fidelity in the characterization process. This aspect is particularly important when considering that many natural drug molecules can exhibit structural similarities, leading to potential misidentification if not analyzed rigorously [43]. Furthermore, the advent of hyphenated techniques such as two-dimensional gas chromatography (GC × GC) coupled with mass spectrometry has expanded the scope of GCMS applications in natural drug research. Such advances allow for greater resolving power and enhanced peak capacity, thereby enabling better separation of complex mixtures seen in herbal extracts [43]. The multidimensional nature of these methods enhances both the qualitative and quantitative analyses, paving the way for detailed profiling of natural products [44].

The integration of advanced chromatography analytical methods, particularly GCMS, has revolutionized the landscape of natural drug molecule development, isolation, and identification. Through its superior separation capabilities, coupled with precise identification and quantification, GCMS not only aids researchers in discovering new therapeutic agents but also ensures the reliability of these findings. The continuous evolution of these methodologies holds promise for further innovations in natural product research, potentially leading to the discovery of novel drugs [45].

3.1 Method development of GC-MS

Gas chromatography-mass spectrometry (GCMS) has become a cornerstone in analytical chemistry, offering unparalleled capabilities in the separation, identification, and quantification of complex mixtures. The development of efficient gas chromatographic columns and advanced temperature programming has played a pivotal role in enhancing these capabilities.

3.1.1 Gas chromatographic columns and temperature programming

Modern gas chromatographic columns are designed with high-performance materials that offer superior separation efficiency. This is crucial for resolving complex mixtures into their individual components. Temperature programming, which involves the precise control of the column temperature during the chromatographic run, optimizes the resolution and sensitivity of the process. By gradually increasing the temperature, compounds with a wide range of volatilities can be effectively separated, leading to sharper peaks and better quantification [46].

3.1.2 Advances in mass spectrometry technology

The mass spectrometry component of GC-MS has seen significant advancements as well. Key developments in electron impact (EI) and chemical ionization (CI) techniques have expanded the range of compounds that can be accurately identified and quantified. EI is particularly effective for producing reproducible fragmentation patterns that are useful for compound identification, while CI offers softer ionization, reducing fragmentation and making it easier to identify molecular ions [47].

3.1.3 Automated sample injection systems

The integration of automated sample injection systems has streamlined the sample preparation process. These systems minimize human error, enhance reproducibility, and increase throughput by allowing multiple samples to be processed sequentially. This automation is especially beneficial in high-throughput environments such as pharmaceutical development and environmental monitoring [48].

3.1.4 Sophisticated data analysis software

The interpretation of complex chromatographic and mass spectrometric data has also been revolutionized by advanced data analysis software. These programs offer robust algorithms for peak detection, integration, and compound identification, significantly speeding up the data processing pipeline. They also provide tools for quantification and statistical analysis, making it easier to derive meaningful insights from the data [49].

3.2 Liquid chromatography-mass spectrometry (LC-MS)

LC-MS has emerged as a pivotal analytical tool in the field of pharmaceutical sciences, particularly in the development, isolation, and identification of natural drug molecules. The combination of liquid chromatography (LC) and mass spectrometry (MS) provides a powerful platform for analyzing complex biological matrices and addressing the challenges associated with natural product chemistry [50]. Advanced chromatography analytical methods, including high-performance liquid chromatography (HPLC) and ultra-high-performance liquid chromatography (UHPLC), play a crucial role in the separation of chemical constituents from natural sources. These techniques are characterized by their ability to resolve closely related compounds efficiently, which is essential when dealing with the intricate profiles of natural products. By optimizing various parameters such as column selection, mobile phase composition, and gradient elution profiles, researchers can achieve optimal separation of target molecules from a background of other substances [51].

Moreover, the development of novel stationary phases and advances in chromatography technology have enabled enhanced selectivity and sensitivity for natural drug molecules. For instance, the introduction of chiral columns allows for the resolution of enantiomers, which is particularly important in drug development due to the differing biological activities of these compounds. Following isolation, the identification of natural drug molecules is critically facilitated by mass spectrometry. MS is an essential component of the LC-MS system, enabling the determination of molecular weight and structural characterization of compounds. Techniques such as electrospray ionization (ESI) and matrix-assisted laser desorption/ionization (MALDI) ionization allow for the analysis of a wide range of polar and non-polar substances, expanding the applicability of LC-MS across various drug classes [52].

Data acquisition and analysis in LC-MS have significantly progressed with the advent of software solutions capable of utilizing sophisticated algorithms for spectral analysis and deconvolution. These tools enhance the identification of compounds by comparing acquired spectra with known databases, streamlining the process of recognizing unknown natural drug molecules. Furthermore, tandem mass spectrometry (MS/MS) offers enhanced specificity and sensitivity, facilitating the elucidation of complex structures and thus allowing for more confident identification of natural products [53]. The integration of advanced chromatography analytical methods with LC-MS offers unparalleled advantages in the field of drug discovery and development. By enabling effective separation, enhanced sensitivity, and detailed structural elucidation, LC-MS serves as a cornerstone in the identification of bioactive compounds sourced from natural products. These advancements are increasingly pivotal, as there is a growing recognition of the therapeutic potential of natural drug molecules in pharmaceutical applications [54].

The development, isolation, and identification of natural drug molecules benefit significantly from advanced chromatography analytical methods, particularly when combined with mass spectrometry. This synergy not only enhances the efficiency and effectiveness of natural product analysis but also plays a vital role in the exploration and validation of potential therapeutic agents from nature. As technology continues to evolve, the role of LC-MS in pharmaceutical sciences will likely expand, driving innovations in the discovery and development of new natural drugs [55].

3.3 Method development of LC-MS

The landscape of liquid chromatography-mass spectrometry (LCMS) has witnessed remarkable advancements, particularly in the realm of ionization techniques. Two key innovations, electrospray ionization (ESI) and atmospheric pressure chemical ionization (APCI), have been pivotal in revolutionizing mass spectrometry (MS) [56].

3.3.1 Electrospray ionization (ESI)

ESI has transformed the analysis of large biomolecules, such as proteins and peptides. This technique generates ions from macromolecules in solution by applying a high voltage to a liquid sample, producing a fine aerosol. The charged droplets undergo desolvation, resulting in the formation of gas-phase ions. ESI is particularly advantageous for analyzing polar and high-mass compounds, making it indispensable in proteomics and metabolomics [57].

3.3.2 Atmospheric pressure chemical ionization (APCI)

APCI, on the other hand, excels in the ionization of less polar compounds. This technique involves the nebulization of the sample into a corona discharge region, where ionization occurs at atmospheric pressure. APCI is particularly suitable for small to medium-sized organic molecules, including lipids and non-polar pharmaceuticals. Its robustness and ability to handle complex matrices make it a valuable tool in various analytical applications [58].

3.3.3 Tandem mass spectrometry (MS/MS)

The integration of tandem mass spectrometry (MS/MS) has further elevated the capabilities of LCMS. MS/MS involves multiple stages of mass analysis, allowing for the selection and fragmentation of specific ions. This process yields detailed structural information, enabling the identification and quantification of components within complex mixtures with unparalleled accuracy. The ability to perform targeted and untargeted analyses has made MS/MS an essential technique in areas such as drug development and environmental monitoring [59].

3.3.4 Advanced data acquisition and analysis

The evolution of sophisticated software for data acquisition and analysis has been a game-changer in mass spectrometry. These software solutions facilitate the interpretation of complex datasets, offering improved precision and efficiency. Automated peak detection, quantification algorithms, and advanced spectral deconvolution are just a few examples of how software enhancements are streamlining the analytical workflow. As a result, researchers can extract meaningful insights from their data more effectively, driving scientific discovery forward [60].

4. Supercritical fluid chromatography (SFC)

SFC is an advanced analytical technique that has gained prominence in the field of chromatography, particularly for the development, isolation, and identification of natural drug molecules. This method utilizes a supercritical fluid—commonly carbon dioxide—which exhibits unique properties that lie between those of gases and liquids. The ability to manipulate the density and viscosity of supercritical fluids allows SFC to achieve high-resolution separations and enhanced mass transfer, making it especially suited for the analysis of complex mixtures [61].

One of the pivotal advantages of SFC is its efficiency in separating chiral compounds, which is a common requirement in drug development. Many natural drug molecules possess chirality, which can influence their pharmacological profiles. Traditional methods, such as liquid chromatography, often require lengthy separation times and extensive use of organic solvents. In contrast, SFC employs supercritical CO_2, which can be modified with small amounts of organic solvents to optimize solvation properties, thereby facilitating faster analysis and more environmentally friendly practices [62]. The rapid analysis time associated with SFC, coupled with its excellent peak shape and resolution, makes it an ideal choice for high-throughput screening in the pharmaceutical industry.

Isolation of natural drug molecules through SFC is another significant application, particularly in the extraction of bioactive compounds from plant materials. The selective solvation properties of supercritical fluids permit effective extraction of such compounds while minimizing the co-extraction of unwanted matrix constituents. This is a critical consideration when working with complex plant matrices, where the presence of undesired substances can obscure the analysis of the targeted biomolecules [63]. Researchers have shown that SFC can be employed not only for the extraction of these compounds but also for their subsequent analysis. This dual-functionality underscores the method's versatility and efficiency in obtaining pure samples for further characterization.

The identification of natural drug molecules is another area where SFC excels. Coupled with mass spectrometry (MS), SFC can provide structural information about the compounds being analyzed. This combination enhances the possibilities for qualitative analysis and quantitative measurement. The sensitivity and specificity provided by mass spectrometry allow for the elucidation of complex structures and the determination of molecular weights, which is crucial for identifying unknown substances in natural products [64]. The adaptability of SFC methods to various detection techniques further amplifies its capability for identifying a wide range of compounds.

Advanced chromatography analytical methods such as supercritical fluid chromatography are pushing the boundaries of traditional analytical techniques in the development, isolation, and identification of natural drug molecules. The benefits of improved efficiency, reduced solvent usage, and enhanced resolution make SFC a vital tool in drug discovery and analysis. As the pharmaceutical industry continues to evolve toward more sustainable practices, SFC offers an innovative approach to addressing many of the contemporary challenges faced in the isolation and identification of biologically active compounds.

4.1 Method development of supercritical fluid chromatography (SFC)

Supercritical fluid chromatography (SFC) has emerged as a cutting-edge technique in the realm of analytical chemistry due to its efficiency, environmental benefits, and superior separation capabilities. This sophisticated method leverages supercritical fluids—substances at a temperature and pressure above their critical point—primarily carbon dioxide (CO_2), to perform chromatographic separations [65].

4.1.1 Enhanced precision in pressure and temperature control

Advances in supercritical fluid technology have significantly improved the precision with which pressure and temperature can be controlled. This precision is crucial because the solubility and selectivity of compounds in supercritical fluids are highly sensitive to changes in these parameters. This development allows for more consistent and reproducible results in analytical procedures [66].

4.1.2 Environmental friendliness

SFC is considered a green technology due to its use of CO_2 as a mobile phase, which is less harmful to the environment compared to the organic solvents used in traditional liquid chromatography. Moreover, CO_2 can be recycled and reused in the system, further reducing environmental impact [67].

4.1.3 Development of chiral stationary phases

One of the significant advancements in SFC is the creation of chiral stationary phases. These phases are essential for separating enantiomers—molecules that are non-superimposable mirror images of each other. Enantiomers can exhibit vastly different biological activities, making their separation crucial in pharmaceutical development. Chiral stationary phases have enabled more effective and efficient resolution of these compounds, enhancing the drug development process [68].

4.1.4 Integration with mass spectrometry

The coupling of SFC with mass spectrometry (MS) has been a game-changer. This integration combines the separation power of SFC with the identification capabilities of MS, allowing for the rapid and accurate analysis of complex mixtures. This powerful combination is particularly beneficial in fields like pharmaceuticals, where the accurate identification of compounds is essential [69].

Parameter	HPLC	UPLC	GC-MS	LC-MS	SFC
Sample Preparation	Extraction	Extraction	Extraction and chemical derivatization	Extraction	Extraction
Sample Volume	5 μL	2 μL	Less than 1 μL (split)	1 μL (splitless)	50–200 μL
Chromato-graphic Separation	High resolution	Higher resolution	High resolution	Medium resolution	High resolution
Column Temperature	30°C	65°C	150–300°C	50–80°C	Variable (often near room temp)
Mobile Phase	Liquid	Liquid	Gas	Liquid	Supercritical fluid
Detection Method	UV-Vis, Fluorescence	UV-Vis, Fluorescence	Mass Spectrometry	Mass Spectrometry	UV, Fluorescence, Mass Spectrometry
Separation Mechanism	Liquid-liquid partitioning	Liquid-liquid partitioning	Volatile compounds based on boiling point	Liquid-liquid partitioning	Supercritical fluid properties
Speed of Analysis	Moderate	Fast	Moderate to fast	Fast	Fast
Sensitivity	Moderate	High	Moderate to high	High	High
Applications	Pharmaceutical, environmental, food analysis	Pharmaceutical, natural products	Environmental, food, forensic	Pharmaceutical, proteomics	Natural products, food analysis
Cost of Equipment	Moderate	Higher than HPLC	Moderate to high	High	High

Table 2.
Comparative study of parameters used in HPLC, UPLC, GCMS and LCMS technique.

These are all powerful analytical chromatographic techniques used in various fields such as pharmaceuticals, environmental analysis, and biochemistry. Each technique has its own set of parameters that influence its performance and application. Comparative study of parameters used in HPLC, UPLC, GC-MS, and LC-MS techniques are illustrated in **Table 2**.

5. Conclusion

The complexity and diversity of herbal components present significant analytical challenges, necessitating the use of advanced chromatographic techniques such as HPLC, UPLC, LC-MS, GC-MS, and supercritical fluid chromatography (SFC). HPLC is a powerful analytical technique used to separate, identify, and quantify each component in a mixture. It operates by having different components of the mixture travel at different speeds through a stationary phase, leading to their separation. UPLC takes this a step further, providing even more precise and rapid analyses. This makes UPLC ideal for high-throughput environments where speed and accuracy are paramount. LC-MS is particularly useful for identifying unknown compounds and studying complex mixtures. It offers unparalleled sensitivity and specificity, making it a valuable tool for detailed molecular analysis. GC-MS is highly effective for analyzing volatile compounds such as essential oils and fragrances, which are commonly found in herbal products. Its efficacy in separating and identifying volatile substances makes it indispensable in the fragrance and flavor industries. SFC is particularly well-suited for the analysis of non-volatile and thermally labile compounds. This makes it a valuable tool in the analysis of herbal components that might degrade under the conditions used in other chromatographic techniques. In summary, the complexity of herbal components demands the use of advanced chromatographic techniques to ensure accurate and comprehensive analysis. Each technique—HPLC, UPLC, LC-MS, GC-MS, and SFC—offers distinct advantages, making them indispensable tools in the field of herbal analysis. These advancements ensure that researchers can obtain accurate, reliable, and comprehensive data, ultimately leading to a better understanding of the complex mixtures present in herbal components.

Acknowledgements

Author acknowledges the use of Chatgpt AI tools for language polishing of the manuscript.

Conflict of interest

The authors declare no conflict of interest.

Author details

Paranthaman Subash[1], K.K. Senthil Kumar[1], Kareti Srinivasa Rao[2] and Sulekha Khute[1*]

1 Department of Pharmacy, Sri Shanmugha College of Pharmacy, Salem, Tamil Nadu, India

2 Department of Pharmacy, Indira Gandhi National Tribal University, Madhya Pradesh, India

*Address all correspondence to: sulekhakhunte@gmail.com; sulekhakhute.pharmacy@shanmugha.edu.in

References

[1] Asma ST, Acaroz U, Imre K, Morar A, Shah SRA, Hussain SZ, et al. Natural products/bioactive compounds as a source of anticancer drugs. Cancers. 2022;**14**(24):6203

[2] Yuan H, Ma Q, Ye L, Piao G. The traditional medicine and modern medicine from natural products. Molecules. 2016;**21**(5):559

[3] Shrivastava SS, Kharabe PM. Ancient roots of modern medicines; their prospects and promises. Pharmacognosy Reviews. 2021;**15**(29):21-30

[4] Lee J, Chung EK, Kang SW, Lee HJ, Rhie SJ. Quantification of teicoplanin using the HPLC-UV method for clinical applications in critically ill patients in Korea. Pharmaceutics. 2021;**13**(4):572

[5] Kumar A, Jat S, Kumar P, Gulbake A. QbD-driven RP-HPLC method for novel chemo-herbal combination, in-silico, force degradation studies, and characterization of dual drug-loaded polymeric and lipidic nanocarriers. Future Journal of Pharmaceutical Sciences. 2023;**9**(1):110

[6] Saha S, Singh D, Rangari S, Negi L, Banerjee T, Dash S, et al. Extraction optimization of neem bioactives from neem seed kernel by ultrasonic assisted extraction and profiling by UPLC-QTOF-ESI-MS. Sustainable Chemistry and Pharmacy. 2022;**29**:100747

[7] Wang S, Wang B, Dong K, Li J, Li Y, Sun H. Identification and quantification of anthocyanins of 62 blueberry cultivars via UPLC-MS. Biotechnology & Biotechnological Equipment. 2022;**36**(1):587-597

[8] Zhang S, Yang M, Li Y, Wang Y, Lu Y, Cheng Z, et al. Occurrence, distribution, and human exposure of emerging liquid crystal monomers (LC-MS) in indoor and outdoor dust: A nationwide study. Environment International. 2022;**164**:107295

[9] Kumar RN, Prasanth D, Midthuri PG, Ahmad SF, Badarinath AV, Karumanchi SK, et al. Unveiling the cardioprotective power: Liquid chromatography–mass spectrometry (LC–MS)-analyzed Neolamarckia cadamba (Roxb.) Bosser leaf ethanolic extract against myocardial infarction in rats and in silico support analysis. Plants. 2023;**12**(21):3722

[10] Karan K, Singh D, Singh PK, Bharati B, Singh TP, Berndtsson R. Implications of future climate change on crop and irrigation water requirements in a semi-arid river basin using CMIP6 GC-MS. Journal of Arid Land. 2022;**14**(11):1234-1257

[11] Wang L, Li Y, Li M, Li L, Liu F, Li Liu D, et al. Projection of precipitation extremes in China's mainland based on the statistical downscaled data from 27 GC-MS in CMIP6. Atmospheric Research. 2022;**280**:106462

[12] Ganzera M, Zwerger M. Analysis of natural products by SFC–applications from 2015 to 2021. TrAC Trends in Analytical Chemistry. 2021;**145**:116463

[13] Wang X, Xing H, Song F, Luo S, Dai P. Dynamic multicast-oriented virtual network function placement with SFC request prediction. In: 2022 14th International Conference on Communication Software and Networks (ICCSN). China: IEEE; 2022. pp. 81-88

[14] Siddique IM. Unveiling the power of high-performance liquid

chromatography: Techniques, applications, and innovations. India: European Journal of Advances in Engineering and Technology. 2021;**8**(9):79-84

[15] Fernandes GFDS, Salgado HRN, Santos JLD. A critical review of HPLC-based analytical methods for quantification of linezolid. Critical Reviews in Analytical Chemistry. 2020;**50**(3):196-211

[16] Kumar SD, Kumar DH. Importance of RP-HPLC in analytical method development: A review. International Journal of Pharmaceutical Sciences and Research. 2012;**3**(12):4626

[17] Stoll DR, Zhang K, Staples GO, Beck A. Recent advances in two-dimensional liquid chromatography for the characterization of monoclonal antibodies and other therapeutic proteins. Advances in Chromatography. 2019;**1**:29-70

[18] Gaudêncio SP, Bayram E, Lukić Bilela L, Cueto M, Díaz-Marrero AR, Haznedaroglu BZ, et al. Advanced methods for natural products discovery: Bioactivity screening, dereplication, metabolomics profiling, genomic sequencing, databases and informatic tools, and structure elucidation. Marine Drugs. 2023;**21**(5):308

[19] Patel TS, LoBrutto R. Role of HPLC during formulation development. In: HPLC for Pharmaceutical Scientists. United States of America: John Wiley & Sons, Inc.; 2007. pp. 679-734

[20] Young CS, Weigand RJ. An efficient approach to column selection in HPLC method development. LC GC North America. 2002;**20**(5):464-473

[21] Heinisch S, Rocca JL. Effect of mobile phase composition, pH and buffer type on the retention of ionizable compounds in reversed-phase liquid chromatography: Application to method development. Journal of Chromatography A. 2004;**1048**(2):183-193

[22] Abba SI, Usman AG, Selin I. Simulation for response surface in the HPLC optimization method development using artificial intelligence models: A data-driven approach. Chemometrics and Intelligent Laboratory Systems. 2020;**201**:104007

[23] Bhardwaj SK, Dwivedia K, Agarwala DD. A review: HPLC method development and validation. International Journal of Analytical and Bioanalytical Chemistry. 2015;**5**(4):76-81

[24] Tavakoli N, Varshosaz J, Dorkoosh F, Zargarzadeh MR. Development and validation of a simple HPLC method for simultaneous in vitro determination of amoxicillin and metronidazole at single wavelength. Journal of Pharmaceutical and Biomedical Analysis. 2007;**43**(1):325-329

[25] Kozlowski ES, Dalterio RA. Analyte solvent and injection volume as variables affecting method development in semipreparative reversed-phase liquid chromatography. Journal of Separation Science. 2007;**30**(14):2286-2292

[26] Wiczling P, Markuszewski MJ, Kaliszan R. Determination of p K a by pH gradient reversed-phase HPLC. Analytical Chemistry. 2004;**76**(11): 3069-3077

[27] Russo R, Guillarme D, Rudaz S, Bicchi C, Veuthey JL. Evaluation of the coupling between ultra performance liquid chromatography and evaporative light scattering detector for selected phytochemical applications. Journal of Separation Science. 2008;**31**(13):2377-2387

[28] Liu Y, Dong T, Yan K, Wang Z, He G, Zhang Y, et al. A comprehensive and high-throughput screening method for multiple prohibited substances by UPLC-QE plus-HRMS and HPLC-QQQ-MS in human urine for doping control. Analytical Methods. 2023;**15**(26):3206-3224

[29] Lee M, Song JH, Choi EJ, Yun YR, Lee KW, Chang JY. UPLC-QTOF-MS/MS and GC-MS characterization of phytochemicals in vegetable juice fermented using lactic acid bacteria from kimchi and their antioxidant potential. Antioxidants. 2021;**10**(11):1761

[30] Stuart KA, Welsh K, Walker MC, Edrada-Ebel R. Metabolomic tools used in marine natural product drug discovery. Expert Opinion on Drug Discovery. 2020;**15**(4):499-522

[31] Wolfender JL, Litaudon M, Touboul D, Queiroz EF. Innovative omics-based approaches for prioritisation and targeted isolation of natural products–new strategies for drug discovery. Natural Product Reports. 2019;**36**(6):855-868

[32] Rainville PD, Simeone JL, McCarthy SM, Smith NW, Cowan D, Plumb RS. Investigation of microbore UPLC and nontraditional mobile phase compositions for bioanalytical LC–MS/MS. Bioanalysis. 2012;**4**(11):1287-1297

[33] Sri RS, Sri KB, Mounika C. A review on comparative study of HPLC and UPLC. Research Journal of Pharmacy and Technology. 2020;**13**(3):1570-1574

[34] Zhang Q, Shi Y, Ma L, Yi X, Ruan J. Metabolomic analysis using ultra-performance liquid chromatography-quadrupole-time of flight mass spectrometry (UPLC-Q-TOF MS) uncovers the effects of light intensity and temperature under shading treatments on the metabolites in tea. PLoS One. 2014;**9**(11):e112572

[35] Eswarudu MM, Eswaraiah MC, Kumar KP, Sudhakar K. Ultra performance liquid chromatography (UPLC): A preeminent technique in pharmaceutical analysis. Research Journal of Pharmacy and Technology. 2012;**5**(12):1484-1489

[36] Sánchez-Illana Á, Pérez-Guaita D, Cuesta-García D, Sanjuan-Herráez JD, Vento M, Ruiz-Cerdá JL, et al. Model selection for within-batch effect correction in UPLC-MS metabolomics using quality control-support vector regression. Analytica Chimica Acta. 2018;**1026**:62-68

[37] Szkudzińska K, Smutniak I, Rubaj J, Korol W, Bielecka G. Method validation for determination of amino acids in feed by UPLC. Accreditation and Quality Assurance. 2017;**22**:247-252

[38] Gumułka P, Żandarek J, Dąbrowska M, Starek M. UPLC technique in pharmacy—An important tool of the modern analyst. PRO. 2022;**10**(12):2498

[39] Waseem R, Low KH. Advanced analytical techniques for the extraction and characterization of plant-derived essential oils by gas chromatography with mass spectrometry. Journal of Separation Science. 2015;**38**(3):483-501

[40] Ranjan S, Chaitali ROY, Sinha SK. Gas chromatography–mass spectrometry (GC-MS): A comprehensive review of synergistic combinations and their applications in the past two decades. Journal of Analytical Sciences and Applied Biotechnology. 2023;**5**(2):5-2

[41] Koparde AA, Doijad RC, Magdum CS. Natural products in drug discovery. In: Pharmacognosy-Medicinal Plants. London, UK: IntechOpen; 2019

[42] Anthony IG, Brantley MR, Floyd AR, Gaw CA, Solouki T. Improving accuracy and confidence of chemical identification by gas chromatography/vacuum ultraviolet spectroscopy-mass spectrometry: Parallel gas chromatography, vacuum ultraviolet, and mass spectrometry library searches. Analytical Chemistry. 2018;**90**(20):12307-12313

[43] Beale DJ, Pinu FR, Kouremenos KA, Poojary MM, Narayana VK, Boughton BA, et al. Review of recent developments in GC–MS approaches to metabolomics-based research. Metabolomics. 2018;**14**:1-31

[44] Ramakrishna K, Raman NVVSS, Rao KN, Prasad AVSS, Reddy KS. Development and validation of GC–MS method for the determination of methyl methanesulfonate and ethyl methanesulfonate in imatinib mesylate. Journal of Pharmaceutical and Biomedical Analysis. 2008;**46**(4):780-783

[45] Blumberg LM. Temperature-Programmed Gas Chromatography. Germany: John Wiley & Sons; 2010. pp. 41-63

[46] Kopka J. Current challenges and developments in GC–MS based metabolite profiling technology. Journal of Biotechnology. 2006;**124**(1):312-322

[47] Misra BB. Advances in high resolution GC-MS technology: A focus on the application of GC-Orbitrap-MS in metabolomics and exposomics for FAIR practices. Analytical Methods. 2021;**13**(20):2265-2282

[48] Čajka T, Maštovská K, Lehotay SJ, Hajšlová J. Use of automated direct sample introduction with analyte protectants in the GC–MS analysis of pesticide residues. Journal of Separation Science. 2005;**28**(9-10):1048-1060

[49] Behrends V, Tredwell GD, Bundy JG. A software complement to AMDIS for processing GC-MS metabolomic data. Analytical Biochemistry. 2011;**415**(2):206-208

[50] Munjanja BK. Liquid chromatography mass spectrometry (LC-MS). In: Multiresidue Methods for the Analysis of Pesticide Residues in Food. CRC Press; 2017. pp. 197-234

[51] Onghena M, Moliner-Martinez Y, Picó Y, Campíns-Falcó P, Barceló D. Analysis of 18 perfluorinated compounds in river waters: Comparison of high performance liquid chromatography–tandem mass spectrometry, ultra-high-performance liquid chromatography–tandem mass spectrometry and capillary liquid chromatography–mass spectrometry. Journal of Chromatography A. 2012;**1244**:88-97

[52] Aszyk J, Byliński H, Namieśnik J, Kot-Wasik A. Main strategies, analytical trends and challenges in LC-MS and ambient mass spectrometry–based metabolomics. TrAC Trends in Analytical Chemistry. 2018;**108**:278-295

[53] Sullards MC, Liu Y, Chen Y, Merrill AH Jr. Analysis of mammalian sphingolipids by liquid chromatography tandem mass spectrometry (LC-MS/MS) and tissue imaging mass spectrometry (TIMS). Biochimica et Biophysica Acta (BBA)-Molecular and Cell Biology of Lipids. 2011;**1811**(11):838-853

[54] Lee MS. LC/MS Applications in Drug Development. Germany: John Wiley & Sons; 2003. pp. 34-178

[55] Hoke SH II, Morand KL, Greis KD, Baker TR, Harbol KL, Dobson RL. Transformations in pharmaceutical research and development, driven by innovations in multidimensional mass spectrometry-based technologies.

International Journal of Mass Spectrometry. 2001;**212**(1-3):135-196

[56] Kailasa SK, Koduru JR, Park TJ, Wu HF, Lin YC. Progress of electrospray ionization and rapid evaporative ionization mass spectrometric techniques for the broad-range identification of microorganisms. Analyst. 2019;**144**(4):1073-1103

[57] Lam W, Ramanathan R. In electrospray ionization source hydrogen/deuterium exchange LC-MS and LC-MS/MS for characterization of metabolites. Journal of the American Society for Mass Spectrometry. 2002;**13**(4):345-353

[58] Raffaelli A. Atmospheric pressure chemical ionization (APCI): New avenues for an old friend. Advances in LC–MS Instrumentation. 2007;**72**:11-25

[59] Vogeser M, Parhofer KG. Liquid chromatography tandem-mass spectrometry (LC-MS/MS)-technique and applications in endocrinology. Experimental and Clinical Endocrinology & Diabetes. 2007;**115**(09):559-570

[60] Ma S, Chowdhury SK. Data acquisition and data mining techniques for metabolite identification using LC coupled to high-resolution MS. Bioanalysis. 2013;**5**(10):1285-1297

[61] Taguchi K, Fukusaki E, Bamba T. Supercritical fluid chromatography/mass spectrometry in metabolite analysis. Bioanalysis. 2014;**6**(12):1679-1689

[62] Soni P, Pabla G, Verma K, Ganti SS, Bhatia R. Recent applications of supercritical fluid chromatography in modern analysis: Updates from 2017 to 2020. Current Analytical Chemistry. 2021;**17**(6):857-882

[63] Liu LX, Zhang Y, Zhou Y, Li GH, Yang GJ, Feng XS. The application of supercritical fluid chromatography in food quality and food safety: An overview. Critical Reviews in Analytical Chemistry. 2020;**50**(2):136-160

[64] Li P, Wu DR, Yip SH, Sun D, Pawluczyk J, Smith A, et al. Large-scale purification of a deprotected macrocyclic peptide by supercritical fluid chromatography (SFC) integrated with liquid chromatography in discovery chemistry. Journal of Chromatography A. 2024;**1730**:465112

[65] Zhao Y, Woo G, Thomas S, Semin D, Sandra P. Rapid method development for chiral separation in drug discovery using sample pooling and supercritical fluid chromatography–mass spectrometry. Journal of Chromatography A. 2003;**1003**(1-2):157-166

[66] Åsberg D, Enmark M, Samuelsson J, Fornstedt T. Evaluation of co-solvent fraction, pressure and temperature effects in analytical and preparative supercritical fluid chromatography. Journal of Chromatography A. 2014;**1374**:254-260

[67] Wu W, Zhang Y, Wu H, Zhou W, Cheng Y, Li H, et al. Simple, rapid, and environmentally friendly method for the separation of isoflavones using ultra-high performance supercritical fluid chromatography. Journal of Separation Science. 2017;**40**(13):2827-2837

[68] Regalado EL, Welch CJ. Separation of achiral analytes using supercritical fluid chromatography with chiral stationary phases. TrAC Trends in Analytical Chemistry. 2015;**67**:74-81

[69] Sakai M, Hayakawa Y, Funada Y, Ando T, Fukusaki E, Bamba T. Development of a practical online supercritical fluid extraction–supercritical fluid chromatography/mass spectrometry system with an integrated split-flow method. Journal of Chromatography A. 2019;**1592**:161-172

Chapter 5

Microfluidic Detection Technologies and Applications

Qiao Cao and Xiangyu Chen

Abstract

Microfluidic chips can condense an entire analytical detection laboratory into a chip of just a few square centimeters, achieving miniaturization, integration, automation, and high-sensitivity. The main microfluidic detection technologies include electrochemical detection technology, capacitively coupled contactless conductivity detection technology, and photoelectric detection technology, which are now widely used in environmental monitoring, food safety testing, drug testing, disease diagnosis, and agricultural detection. In this chapter, we summarize the typical analytical methods integrated onto microfluidic platforms. Applications of these microfluidic analytical methods on environment, food safety, biomedicine, and other fields also have been discussed. At last, the challenges and future directions about microfluidics-based analysis development have been remarked.

Keywords: microfluidic chip, detection technologies, environment monitoring, food safety, biomedicine

1. Introduction

The micro-total analysis system (lab-on-a-chip, LOC), first proposed by Widmer and Manz in 1990, has developed rapidly in recent decades as a new technology platform [1]. In 2001, the journal of the same name "Lab on a Chip" was founded and quickly became a mainstream publication, leading to the development of related research worldwide. Subsequently, the LOC was named "one of the 15 most important inventions affecting the future of mankind" by Forbes magazine in 2003 and "one of the seven technologies that will change the future" by Business magazine in 2004. Nature also published a "Laboratory on a chip" album in 2006, which made a comprehensive description of the development and application prospects of the lab on a chip and pointed out that LOC had great competitiveness to become "the technology of the century" [2–4].

As an important supporting technology of LOC, microfluidic technology is a new technology platform that uses microtubes (tens to hundreds of micrometers in size) to process or manipulate tiny fluids (nanoliters to liters in volume). Microfluidic chip is the core of microfluidic technology, on which researchers manipulate the fluid flow and accelerate the reaction process with the help of surface tension, fluid resistance, and energy dissipation [5–7]. Different from the large equipment required by traditional methods, microfluidic devices have the advantages of miniaturization and integration. Compared with traditional methods, microfluidic technology can overcome the shortcomings of long detection time and complex pre-processing to a certain extent, and meet the requirements of real-time on-site detection, due to its advantages of small size, low cost, good compatibility, and easy integration.

Due to the advantages of time-saving, reagent-saving and high efficiency, microfluidic technology has attracted lots of attention from researchers and has been developed explosively in the past few decades. On the one hand, in order to obtain low-cost microfluidic chips, researchers have made great progress in the development of new materials and the manufacture of chips. Novel chip materials include polymethylmethacrylate (PMMA) [8], polydimethylsiloxane (PDMS) [9], and paper [10] which are suitable for mass production; meanwhile, 3D printing technology and low-cost bonding methods [11–13] have greatly reduced chip processing costs. On the other hand, microfluidics has developed many detection methods, such as spectroscopy, mass spectrometry, and electrochemistry, and were widely applied in food safety, environmental monitoring, disease diagnosis, biological analysis, and other fields [14–18].

In this review, we summarize the latest progress of microfluidic, making a brief description of microfluidic from the aspects of working principle, detection method, and application. In addition, we present an outlook on the challenges and application prospects of microfluidic technology in the future (**Figure 1**).

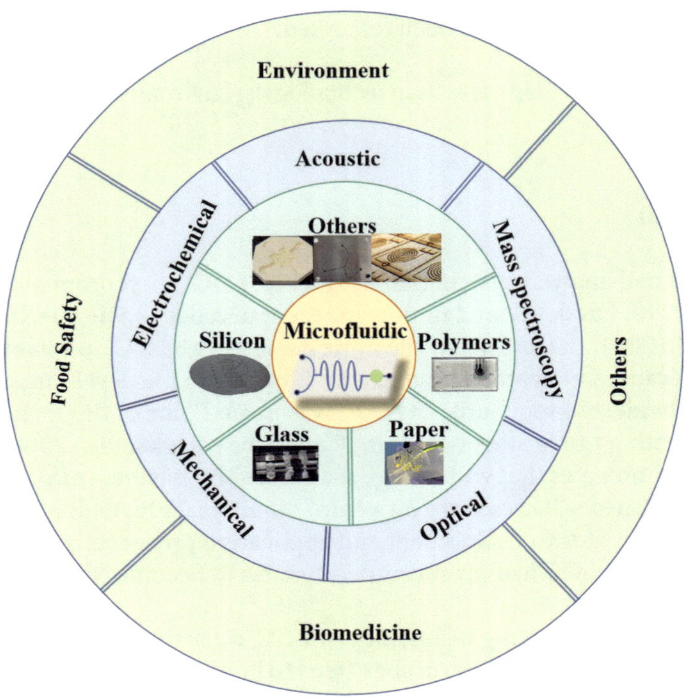

Figure 1.
Summary diagram of applications based on microfluidic technology.

2. The principle of microfluidics

Microfluidics is the science and technology involved in systems that use microchannels to process or manipulate tiny fluids. Based on the unique properties of fluids in microscale environments, such as laminar flow and droplets, microfluidic enables a range of micromachining and micromanipulation that is difficult to achieve with conventional methods. The experimental platform integrating microfluidic drive, control, reaction, and detection is called microfluidic system. A microfluidic system is usually composed of fluid drive subsystem, process monitoring and control subsystem, microfluidic chip, and detection and analysis subsystem (**Figure 2**). Fluid drive subsystem, as the name suggests, is the part that provides driving force for the fluid to move forward in the system, usually composed of microfluidic drive pumps. Commonly used drive pumps include pressure pumps, injection pumps, and peristaltic pumps. In addition, other driving forces such as electro-osmosis, capillary force, and gravity can also drive the fluid forward. The process monitoring and control subsystem is usually composed of flow sensors and valves, which is used to achieve a variety of fluid control such as flow feedback control, sequential injection, circulation injection, and volume quantification. Common fluid control methods are electroosmotic control and microvalve control, including passive valve control and active valve control. Microfluidic chip, as the core component of microfluidic system, is a device for processing tiny fluid traces. These microfluidic chips made of silicon, glass, polymer materials, and paper have microchannels inside and exits/entrances connected with the outside, which can integrate the process of sample preparation, pretreatment, separation, and detection and promote the miniaturization, integration, and automation of the whole process of sample detection. The detection and analysis subsystem are used to observe, record, and analyze the reaction on the microfluidic chip and detect and amplify the signal generated by the microfluidic chip, usually with the help of high-precision equipment such as a high-speed CCD, microscope, or spectrometer. At present, the detection of microfluidic chips mainly uses spectrum, mass spectrometry, and electrochemical detectors. The entire working process of the microfluidic system is as follows. First, under the control of the pump and valve, the reagents and samples enter the microfluidic chip through the inlet to react. The signals caused by the reaction are then detected by a detector. Finally, the waste liquids are discharged through the control of the pump and valve.

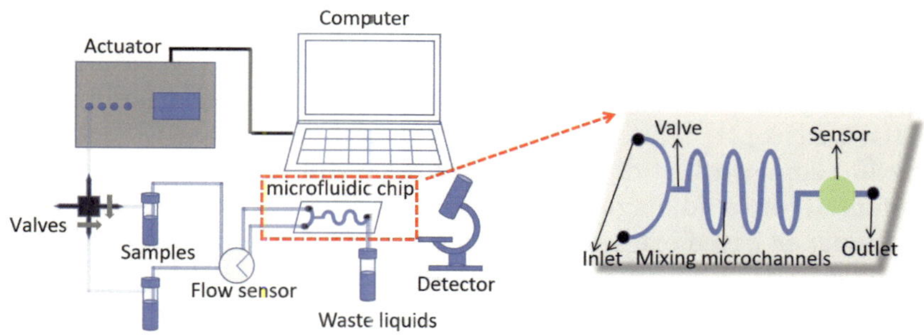

Figure 2.
The composition of microfluidic system and the structure of microfluidic chip.

3. Detection method

3.1 Optical detection principles

Optical detection methods are widely applied in microfluidic chips, primarily utilizing the interaction between light and matter to detect the characteristic parameters of samples. These methods are typically highly sensitive and non-invasive, making them suitable for analyzing various biological, chemical, and physical samples. The principles of optical detection include absorption spectroscopy, fluorescence detection, and surface plasmon resonance (SPR).

3.1.1 Absorption spectroscopy

Absorption spectroscopy and colorimetric analysis techniques have significantly advanced in microfluidic and paper-based platforms, particularly for metal ion detection and rapid onsite analysis. Sasaki et al. proposed an absorption-based colorimetric detection method for nickel (II) ions, based on the phase separation of thermoresponsive magnetic nanoparticles [19]. This method utilized the phase transition of thermoresponsive magnetic nanoparticles upon detecting nickel (II) ions. When these nanoparticles undergo phase transition, they could be separated by magnetic force, leading to a color change in the solution. This enabled rapid detection of nickel ions in environmental samples with high sensitivity and selectivity. Martinez et al. introduced the first microfluidic paper-based analytical device (μPAD), where samples and reagents were transported to the detection area via capillary action [10]. This approach led to the further development of paper-based detection devices modeled after classical qualitative analysis test strips, where samples directly interacted with receptors immobilized on cellulose scaffolds. The patterned paper could be modified for biological assays by applying suitable reagents to the test areas. Additionally, Shen et al. designed a portable colorimetric detection method based on a smartphone, utilizing a color-adaptive algorithm to enhance detection accuracy [20]. This method minimized the impact of device differences on microfluidic platforms, providing a simple and reliable solution for field testing.

3.1.2 Fluorescence detection

Fluorescence detection methods have advanced considerably in microfluidic and μPADs, particularly in biomarker detection and clinical diagnostics. Kong et al. developed a wearable microfluidic device that combines recombinase polymerase amplification (RPA) with fluorescence detection for rapid detection of HIV-1 DNA [21]. This device used body heat to drive the RPA reaction, detecting 100 copies/mL of the target sequence within 24 minutes, demonstrating its potential for rapid diagnostics in resource-limited settings. Xu et al. developed a paper-based origami device for malaria detection through fingertip blood samples [22]. This device contained five panels preloaded with isothermal amplification reagents targeting different malaria species, enabling DNA extraction, purification, and detection on a simple heating plate. The results could be read directly under a handheld UV light. The team later optimized the device by adding a fingertip-driven pump and lateral flow immunochromatography, making it more user-friendly. Field testing in rural Uganda achieved 98% sensitivity. Moreover, Liang et al. introduced an innovative flower-like silver (FLS) enhanced fluorescence/visual bimodal detection platform for ultrasensitive and multiplexed microRNA (miRNA) detection [23]. This platform combined the advantages of visual detection for screening and fluorescence detection for quantitative analysis. It utilized a metal-enhanced fluorescence (MEF) effect, where the FLS layer grown on a paper substrate reduced background fluorescence and enhanced the signal. The biosensor could detect miRNA-210 and miRNA-21 at concentrations as low as 0.03 fM and 0.06 fM, respectively. Additionally, the platform could be reused by replacing the visual substitutes and supplementing the DNA2-CeO_2 complex.

3.1.3 Surface plasmon resonance (SPR)

SPR is a label-free optical detection technique that monitors molecular binding and interactions based on a resonance effect. When combined with microfluidic technology, SPR enables real-time analysis of proteins, antibodies, and other biomolecules, making it suitable for studies of molecular interactions and drug screening. Liu et al. developed a dual-channel SPR sensor based on a microstructure hollow fiber (MHF) for microfluidic applications [24]. The sensor featured an internal sensing channel within the MHF and an external sensing channel outside the MHF. Researchers configured the hollow fiber SPR sensing probe using grinding, polishing, and termination techniques and coated the internal and external sensing channels with a 50 nm gold layer. The sensor detected refractive index changes from 1.333 to 1.385, with a resonance wavelength range of 712–1010 nm and a sensitivity of 5645 nm/RIU for the internal channel, while the external channel shows a range of 618–700 nm and a sensitivity of 1576 nm/RIU. The unique structure of the MHF, featuring air holes, made it well-suited for integration with microfluidic systems. Huang et al. developed a sensitive and cost-effective microfluidic-integrated biosensor based on the localized surface plasmon resonance (LSPR) properties of gold nanoparticles [25]. The biosensor enabled label-free real-time monitoring of biomolecular interactions. It incorporated a novel quadrants detection scheme for continuous measurement of light changes across the nanoparticle-coated sensor surface. Performance and sensitivity of the integrated system were tested using an optimized gold nanoparticle thin film, demonstrating its potential for biosensing applications using antigen/antibody (biotin/anti-biotin) models. Additionally, Yang et al. introduced a compact HSC-SPR structure-based high-sensitivity refractive index (RI) microfluidic chip [26]. This chip consisted of a hollow silica capillary (HSC) segment coated with a thin gold film between two multimode fibers (MMFs). Compared with other optical fiber SPR sensors, this design featured easy fabrication, structural stability, and cost-effectiveness. When measuring refractive index changes from 1.333 to 1.424, the chip exhibited a high sensitivity of 7225.63 nm/RIU.

3.2 Mass spectrometry analysis

Mass spectrometry (MS) detection techniques analyze the mass-to-charge ratio (m/z) of ionized molecules in electric or magnetic fields. Integrated with microfluidic platforms, MS detection enables high-resolution analysis with minimal sample volume, making it a crucial tool for research in drug metabolism, proteomics, and more. Principles of mass spectrometry analysis include electrospray ionization mass spectrometry (ESI-MS) and matrix-assisted laser desorption/ionization mass spectrometry (MALDI-MS).

3.2.1 Electrospray ionization mass spectrometry (ESI-MS)

ESI-MS is a common ionization technique that ionizes sample molecules through electrospray, suitable for analyzing biomolecules in complex samples. Integrated with microfluidic platforms, ESI-MS can efficiently process small samples, achieving precise detection of drug metabolism and protein interactions. Zhang et al. developed an integrated microfluidic system for real-time single-cell lipid analysis using MS [27]. The system combined droplet-based inkjet printing, dielectrophoretic electrodes, and demulsification interfaces to enable high-throughput single-cell encapsulation, manipulation, and MS detection. This method improved single-cell encapsulation efficiency and reduced matrix and oil-phase interference, allowing the study of lipid variations between normal and cancer cells, as well as differences before and after drug treatment. Zhang et al. introduced a paper spray MS method for analyzing droplets generated in gravity-driven microfluidic chips [28]. This method used capillary action to transfer droplets from the chip to a paper substrate for analysis by paper spray ionization. The manually controlled interface was simple, low-cost, and easy to implement. The method also exhibited good tolerance for complex matrices, allowing direct analysis of real biological/chemical micro-reaction samples. As a proof of concept, the researchers analyzed the acetylcholine hydrolysis reaction and combined a flow injection analysis system with a droplet system to generate concentration gradients, obtaining kinetic information of the micro-reaction in a single sample injection.

3.2.2 Matrix-assisted laser desorption/ionization mass spectrometry (MALDI-MS)

MALDI-MS is a technique that ionizes samples by laser excitation of their surfaces. Integrated with microfluidics, MALDI-MS enables high-throughput sample analysis, particularly suitable for the detection and identification of proteins and peptides, making it widely applicable in biomedical research. Nie et al. proposed a new method that couples microchip electrophoresis with matrix-assisted laser desorption/ionization time-of-flight mass spectrometry (MALDI-TOF-MS) using a freezing technique [29]. Specifically, electrophoretic separation was conducted on a microfluidic chip, and the fluid in the microchannel was immediately frozen to preserve the distribution of separated substances. The frozen chip was then freeze-dried and analyzed as a MALDI target sample in a mass spectrometer. This method avoided sample cross-contamination, providing a novel interface for coupling microfluidic electrophoresis with MALDI-MS. Wang et al. developed a surface-modified indium tin oxide (ITO) glass-PDMS hybrid microfluidic chip for screening potential thrombin inhibitors from natural products [30]. The chip integrated fluorescence imaging and MALDI imaging for monitoring and characterizing enzymatic reactions between thrombin and its inhibitors. Fluorescently labeled substrates were immobilized on an ITO glass substrate modified with gold nanoparticles and thiolated-β-cyclodextrin-coated titanium dioxide nanowires. The PDMS microchannel plate was placed over the modified ITO substrate to form a microreactor environment. Fluorescence imaging was used for in situ monitoring of the enzymatic reaction, while MALDI-MS verified and quantified thrombin hydrolysis products to determine enzyme kinetics and the inhibitory activity of selected flavonoids. The results indicated that rutin, kaempferol, and baicalin have significant thrombin inhibitory activity, with IC50 values below 30 μM.

3.3 Electrical detection

Electrical detection methods identify chemical or biological molecules in a sample by measuring changes in electrical signals. This approach offers high sensitivity and rapid response in microfluidic detection, suitable for various applications such as analyzing electrochemical reactions and monitoring cellular properties. The principles of electrical detection include electrochemical detection and impedance spectroscopy.

3.3.1 Electrochemical detection

Electrochemical detection involves the reaction between electrodes and samples, producing measurable changes in electrical signals such as current or voltage. This method can be enhanced by optimizing electrode materials or designing microelectrode arrays on microfluidic platforms, making it ideal for detecting heavy metal ions, glucose, nucleic acids, and other small molecules. Karikalan et al. developed a disposable microfluidic homocysteine sensor chip (MHS chip) for electrochemical detection, designed for detecting homocysteine (Hcy), a critical biomarker for cardiovascular diseases and strokes [31]. The key innovation lied in using copper-doped graphene (Cu-Gr) as a catalytic reaction bed within the microfluidic channel, facilitating the hydrolysis and detection of N-homocysteinylated proteins (N-Hcy-proteins), a form of Hcy that conventional methods struggle to detect. The MHS chip offered a linear dynamic range of 0.01–124 μM and a detection limit of 2.9 nM, making it suitable for clinical applications. Park et al. developed an integrated pumpless microfluidic chip for detecting foodborne pathogens using polymerase chain reaction (PCR) and electrochemical analysis [32]. The chip incorporated a thin-film-based PCR module, an electrode module, and a PDMS-based finger-driven microfluidic module, enabling sample transport, mixing, gene amplification, and quantification without external pumps. The system demonstrated its efficacy using *Escherichia coli* O157 as a target pathogen, achieving a detection limit of 102 colony-forming units (CFU).

3.3.2 Electrical impedance spectroscopy (EIS)

EIS measures changes in the electrical impedance of a sample to analyze its physical properties or concentration variations, making it ideal for cell analysis and pathogen detection, as it can provide label-free information about the sample's electrical characteristics and enable real-time monitoring. Cao et al. proposed a novel microfluidic aspiration-assisted electrical impedance spectroscopy (MAEIS) system for characterizing oocyte hardening [33]. This system used hydrodynamic capture to handle individual oocytes and measured the hardening status of their cumulus layers through impedance spectroscopy. The researchers demonstrated the system using three different mouse oocyte models with varying cumulus layer properties, allowing determination of the Young's modulus of the cumulus layer. This provided a simple, observer-independent, and reliable method for assessing oocyte quality. The team also developed a microfluidic chip with embedded vertically arranged microelectrodes for impedance measurement [34]. The chip was fabricated using a three-layer dry film resist (DFR) lamination process with a central aperture in the ring electrode and an underlying microfluidic adsorption channel for capturing and securing target objects. By performing impedance measurements on polystyrene and zirconia spheres, they observed maximum sensitivity at approximately 2 kHz. The measurement values correlated with the size of the captured spheres, and equivalent circuit models were used to simulate the experimental results.

3.4 Acoustic detection

Acoustic detection methods analyze samples by utilizing interactions between sound waves and samples, often to assess physical properties or biomolecular interactions. Combined with microfluidic technology, acoustic detection allows for non-invasive analysis of small samples, making it suitable for biomedical and environmental monitoring applications. The principles of acoustic detection include acoustic resonance detection and acoustic levitation.

3.4.1 Acoustic resonance detection

Acoustic resonance detection relies on the coupling between sound waves and samples, measuring changes in resonance frequencies to detect sample characteristics. Microfluidic acoustic resonators offer high sensitivity and fast response times for detecting molecular concentration changes in liquid samples. Yiannacou et al. proposed a method for controlling and actively sorting particles within a closed-channel microfluidic chip using bulk acoustic waves and machine learning [35]. They utilized a multi-armed bandit algorithm to control the frequency of a single ultrasound transducer based on real-time visual tracking of particle positions, which allowed for 2D closed-loop manipulation of particles without prior modeling or calibration of the acoustic field. The method was robust to variations in chip and particle properties and could be used to sort multiple particles simultaneously in a microfluidic chip. Liu et al. reviewed the use of surface acoustic wave (SAW) devices in biological applications [36]. SAWs are elastic waves that propagate along the surface of piezoelectric materials and have been widely used across various fields, including biosensing. The review discussed different types of SAWs, such as Rayleigh SAWs, shear-horizontal SAWs (SH-SAW), Love waves (LW-SAW), and Lamb waves, as well as their applications in biological detection and sample manipulation. Rayleigh SAW devices typically serve as actuators for separating, sorting, and mixing biological particles, while other SAW types are mainly used as biosensors to detect biomolecules, cells, viruses, and bacteria.

3.4.2 Acoustic levitation

Acoustic levitation detects the characteristics of micro-particles or cells by forming an acoustic field within a microfluidic channel, suspending them without contact. This method achieves high-sensitivity detection, making it suitable for analyzing cells and micro-particles. Li et al. developed a droplet lab platform that uses microfluidics and acoustic levitation to design complex, emulsion-based multi-chamber artificial cells [37]. This platform allowed for the creation, manipulation, control, and measurement of artificial cells with distributed cores (ACDC droplets). ACDC droplets consisted of water droplets (cores) enclosed by lipid membranes, forming lipid bilayer networks that enabled communication between isolated cores through protein mediation. The microfluidic production and acoustic levitation of ACDC droplets provided a way to control the internal organization, connectivity, and functionality of the compartments, including the remote activation and deactivation of membrane proteins like large mechanosensitive channels (MscL). Jeger-Madiot et al. designed a multi-node acoustofluidic chip capable of precisely controlling the force and position of acoustic traps over a wide frequency range [38]. The authors used a simple one-dimensional (1D) multilayer model to describe the standing wave generated within the chip's cavity, allowing them to create pressure nodes under non-resonant conditions. The transparent chip cavity and broadband ultrasound transducer enabled the measurement of acoustic energy across the entire frequency spectrum using particle image velocimetry (PIV). Results demonstrated that the acoustofluidic chip could maintain a constant trap amplitude by adjusting the frequency, which was applied to manipulate and merge mesenchymal stem cell (MSC) spheroids cultured in acoustic levitation.

3.5 Mechanical detection

Mechanical detection methods rely on the physical contact or deformation between the sample and a sensor to analyze the sample. This label-free detection approach is highly efficient when integrated with microfluidic technology, allowing for precise analysis of the mechanical properties of samples, fluid characteristics, and more. The principles of mechanical detection include microcantilever detection and pressure sensing.

3.5.1 Microcantilever detection

Microcantilever detection involves changes in the deformation or vibration frequency of a microcantilever structure caused by interactions with the sample, allowing the detection of molecular mass or surface stress. This method is used for detecting DNA, proteins, and other biomolecules, offering high sensitivity and easy integration. Tajeddin et al. designed a microfluidic chip with an integrated pH sensor that could be combined with organ-on-chip models for real-time pH monitoring [39]. The sensor employed a biomimetic design that mimiced intracellular pH regulation mechanisms. The chip consisted of a main channel and a reaction chamber where the sample reacted with a buffering system. Free-flow electrophoresis separated ions in the sample, and the sensor uses a microcantilever as the sensing element. By adjusting the cantilever's geometric and mechanical parameters, the sensitivity of the sensor could be tuned. Chien et al. measured the mass changes of individual living cells in a liquid environment using microcantilevers with sub-nanogram weight sensing capabilities [40]. The researchers demonstrated rapid mass measurements within 3 minutes, detecting the mass of polystyrene and metal microbeads as small as 280 ± 95 pg, as well as live single cells. Finite element analysis was used to simulate and optimize the cantilever design and materials, showing that using polymers like SU8 and polyimide could improve the minimum detectable mass by threefold compared to traditional silicon cantilevers. The main advantages of this approach were the ability to study both adherent and non-adherent cells, to reuse cantilevers for multiple measurements, and to maintain a fixed attachment location for cells at the cantilever tip.

3.5.2 Pressure sensing

Pressure sensing measures changes in pressure as samples flow through microfluidic channels, making it ideal for studying cellular mechanics, fluid characteristics, and the physical parameters of micro-particles. It enables precise measurements within small volumes. Zhang et al. developed a novel on-chip micro-pressure sensor for monitoring pressure in microfluidic systems [41]. The main body of the sensor was a PDMS microfluidic chip containing a working fluid channel and a sealed detection channel. Pressure changed in the working fluid channel deforms the PDMS membrane between the two channels, altering the volume of the detection channel. The detection channel was filled with two immiscible fluids, and pressure is monitored by measuring the displacement of the fluid interface. This design minimized cross-contamination between the working and reference fluids, while reducing costs. Du et al. introduced a microfluidic contact lens integrated with microchannels and micropumps, designed for controlled ocular drug delivery [42]. The contact lens utilized a pressure source to trigger drug release, overcoming the limitations of low corneal bioavailability and short retention time associated with traditional eye drops. The microfluidic contact lens was fabricated using photolithography and mold replication techniques, creating a flat microfluidic chip component that was then cast into a curved shape. Experiments demonstrated that controlled liquid release could be achieved through an outlet check valve. The contact lens exhibited good flexibility, light transmittance, and biocompatibility. It could carry different types of drugs in different regions and used the mechanical action of blinking to facilitate drug exchange between the eye and the lens without electronic components, improving safety and stability. Yu et al. developed a highly sensitive microliter blood pressure sensor based on patterned micro-nanostructure arrays [43]. The sensor had an ultra-high sensitivity of 16.71 $mbar^{-1}$, a low-pressure range of 2 mbar, and required less than 1.3 μL of sample volume. It could measure a wide pressure range, from microfluidic applications to large-volume liquid pressures, including blood pressure. The sensor used a patterned micro-ridge-nanopillar array as a bursting microvalve, and its sensitivity was based on specific flow behaviors through the microvalve array. It provided a visible, in situ pressure record without the need for additional electronic equipment or data acquisition circuits.

4. Application of microfluidics

Due to its advantages of small scale, easy integration, and low cost, microfluidic technology has a wide range of applications in various fields, including environment, food safety, biomedicine, chemistry, and so on.

4.1 Environment

Along with the rapid progress of society, the rapid development of industry and human life activities has caused serious environmental pollution, which has a profound impact on human health and ecosystem, and requires timely monitoring of pollutants and nutrients in the environment. Microfluidic technology is an important means of environmental monitoring, which is widely used in the detection of heavy metals, pesticides, nutrients, and microorganisms in environment.

4.1.1 Heavy metals

Certain heavy metals within the safe range are essential for certain physiological functions, but excessive amounts can cause diseases such as cancer [44–46], so a number of microfluidic devices have been developed to detect heavy metals in the environment. Wang et al. designed a heavy metal chemical sensor based on distance readings to simultaneously detect copper, lead, and silver ions in water with high precision [47]. After the water sample was collected by the microfluidic device, the deoxy-ribozyme (DNAzymes), probe modified magnetic microparticles (MMPs), and polystyrene microparticles (PMPs) pre-loaded into the chip were hybridized to form the "MMPs-DNAzyme-PMPs" structure. The target metal ions could split the "MMPs-DNAzyme-PMPs" structure to produce more free PPMs, which was separated using a gravity magnetic separator and accumulated through a nozzle into the capture channel, forming a visual bar whose distance was proportional to the target metal ion concentration. Huang et al. proposed a fluorescence microfluidic technique based on GO and aptamers for the detection of mercury and lead ions [48]. The fluorescence resonance energy transfer (FRET) between the fluorescent dyes and GO was weakened by the interaction of the special aptamers with Hg and Pb, resulting in fluorescence recovery of the dyes. Bai et al. demonstrated a full femtosecond laser processing technique to fabricate a microfluidic chip that applied surface enhanced Raman spectroscopy (SERS) for the detection of Cd [49]. Inside these microfluidic chips were 3D glass microfluidic channels accompanied by 2D periodic metal nanodot arrays, which provide a high SERS device enhancement factor up to 7.3×10^8 and a sensitive real-time SERS detection of Cd at levels as low as 10 ppb. Ma and coworkers have developed an electrochemical microfluidic chip that amplified Pb ion detection signals through the synergistic effect of 3D Ag-rGO-f-$Ni(OH)_2$/NF composites and thermocapillary convection, which could be combined with smartphones and electrochemical workstations to achieve rapid and efficient Pb detection at the ppb level [50]. In addition to heavy metal ions, some toxic anions can also be detected using microfluidic technology. Lace et al. developed a microfluidic system for the detection of arsenic in water using leucomalachite green [51]. The leucomalachite green could be oxidized to malachite green by the iodine released by the reaction of arsenic with potassium iodate in the acidic medium, resulting in a green absorption band (λ_{max} = 617 nm). The linear detection range of this method was 0.3 to 2 mg/L, which would be used for the detection of arsenic in wastewater.

4.1.2 Nutrients

Nutrients are very important for plant and animal growth, lack of which can make organisms grow slowly and excess can cause disease and pollution. Microfluidic sensors detect major nutrients in the environment including nitrates and nitrites, ammonium salts, and phosphates.

Khanfar et al. designed a PMMA microfluidic chip based on Griess principle for colorimetric detection of nitrate in drinking water [52]. The chip had an LOD of 0.0782 ppm for different water samples, including bottled water, and an R_2 value of 0.98. The method showed good repeatability and accuracy, but the conditions needed to be optimized to improve the sensitivity. Vicent et al. firstly reported the deployment of miniaturized lab-on-a-chip within an autonomous underwater vehicle (AUV) to collect nitrate and nitrite data in seawater [53]. The microfluidic device completed a 21-day deployment of accurate nitrate determination with LOD as low as 0.1 μM (R = 0.98, n = 11) and power consumption as low as 1.5 W. The device's low reagent consumption allowed it to perform 400 dives and 4000 sample measurements. However, this device had the disadvantage of not being able to determine nitrate or nitrite alone. Ali et al. reported an EIS microfluidic device based on a porous graphene foam scaffold modified with nitrate reductase for rapid determination of nitrate in water, showing high sensitivity and selectivity [54]. Li et al. prepared a high-pressure-compatible ion chromatography microfluidic chip with micromilling and laser-based bonding and analyzed standard anionic solution and tap water with an integrated five-electrode conductivity detector (Cl^-, NO_3^-, SO_4^{2-}, and F^-) [55]. Common anions could be separated within 8 minutes with a relative deviation of less than 10% in quantified concentration compared to commercial IC-8286 systems. In addition, paper-based microfluidics have been developed to detect nitrates and nitrites. Mesquita's group developed two novel microfluidic μPADs for quantitative analysis of nitrite and nitrate in human saliva [56]. These μPADs consisted of an overhead layer (9.5 mm diameter Whatman 1 filter paper) and a Griess reagent layer (9.5 mm diameter Whatman 50 filter paper), which were processed and analyzed with the help of standard scanners (Canon LiDE 120) and ImageJ. These μPADs were sample- and user-friendly, eliminating the need for pretreatment and specialized technical requirements.

Ammonia nitrogen is also one of the essential nutrients for biological growth. In recent years, scientists have also developed some microfluidic chips for ammonia nitrogen detection. Breadmore' group developed a microfluidic system based on an improved Berthelot reaction for the colorimetric determination of ammonium in natural water bodies [57]. The system was integrated by a multi-material 3D-printed microfluidic heating reactor and a flow injection analysis system, using Joule heating to accelerate the colorimetric reaction rate to shorten the ammonium analysis time, which was five times faster than the reported method. The detection limit sum of 0.15 mg/L and the linear response of 0.5–5 mg/L ($R^2 = 0.996$) were also compelling advantages of this method. Gallardo-Gonzaleaz et al. designed an in situ real-time electrochemical microfluidic system for the detection of ammonia nitrogen in flowing water [58]. It consisted of a PDMS microfluidic substrate and a fully integrated chemical sensing platform including four working microelectrodes, two reference electrodes, and a counter microelectrode. The system had a low detection limit of 10^{-5} M, a short reaction time of 10–12 s, and a good stability that the sewage immersion of 15 minutes had no influence on normal use, which indicting its good application prospect in aquaculture, river sewage, and seawater monitoring. Cho et al. developed a colorimetric paper-based microfluidic for the detection of ammonia nitrogen [59]. The determination of ammonia was based on an improved Berthelot reaction by selectively forming blue indophenol dye on porous paper. This paper-based sensor had good cross-selectivity, but the detection limit of 10 mg/L hindered its application in low concentration nitrogen detection.

The microfluidic chip design of phosphate is mainly based on spectrophotometric method. Clinton-Bailey's research team improved the power consumption and analytical performance on the basis of commercial phosphate sensors such as the HydroCycle-P (Sea-bird Coastal, Washington State, U.S.A.) and the Wiz Probe (Systea S.p.A. Analytical Technologies, Anagni, Italy) by adding the dispersant polyethylpyrrolidone [60]. They developed a novel in situ microfluidic phosphate sensor with a mean power consumption of 1.8 W, a detection limit of 40 nM, a dynamic range of 0.14–10 μM, and an infield accuracy of 4 ± 4.5%. In order to reduce the difficulty in manufacturing long-channel microfluidic devices caused by the use of complex optics, Luy et al. designed a novel long-path absorbance microfluidic cell embedded with black PMMA on a transparent PMMA substrate as an isolated optical channel [61]. The Z-shaped microchannel structure was designed to form a complete optical window, allowing light coupling into the microchannel. The V-groove prisms integrated seamlessly with the printed circuit board avoiding the use for additional adhesives. This long path absorbance cell did not increase the sample consumption, even the longest 50 mm optical path only used 12 uL of sample with a phosphate detection limit reduced to 40 nM. For the purpose of long-term monitoring of phosphate in natural waters, Grand et al. improved a phosphate analytic device that was not affected by salinity, turbidity, and CDOM when used at room temperature (5–30°C) by eliminating optical interference of salinity and color through optimization [62]. They successfully deployed a month-long accurate hourly monitoring of phosphate in estuarine and coastal environment with a detection limit of 30 nM.

4.1.3 Other pollutants

Other pollutants in the environment include Per- and polyfluoroalkyl substances (PFAS), volatile organic compounds (VOCs), and nitrogen oxides, which pose a serious threat to the ecology and health of life. Fortunately, scientific research in these areas has also achieved some gratifying results.

Because of its stability and durability, PFAS are widely used as a variety of coatings, packaging, etc., but their non-degradable characteristics make them accumulate in the environment and human body inducing health hazards, so their monitoring is also crucial. Cheng et al. developed an electrochemical microfluidic detection platform for perfluorooctane sulfonate (PFOS) based on metal-organic frameworks (MOF) with a Cr central as capture probes [63]. PFOS molecules were adsorbed by capture probes embedded in microfluidic channels and detected using interdigitated microelectrode array located on either side of the channel. The detection limit was as low as 0.5 ng/L which could be comparable to quantitation limits achieved by the most advanced in situ techniques. Breshears et al. used the effects of competitive interactions between perfluorinated-carbon alkyl chains (PFOA), cellulose fibers, and other reagents on surface tension at the wetting front and capillary flow rate to construct a unique paper-based microfluidic chip [64]. The paper-based chip had detection limits of 10 ag/μL for deionized water and 1 fg/μL for treated wastewater and perfectly distinguished PFOA from anionic sodium dodecyl sulfate (SDS), non-ionic Tween 20, and cationic cetrimonium bromide (CTAB) at 100% accuracy.

There is also a class of pollutants in the environment, which are gas pollutants mainly represented by VOCs and nitrogen oxides, causing diseases such as respiratory system and climate change such as acid rain. Guo et al. developed a microfluidic colorimetric chip based on 4-aminohydrazine-5-mercapto-1,2,4-triazole (AHMT) method for gaseous formaldehyde analysis [65]. The chip fabricated by PDMS had two reagent reservoirs and one reaction reservoir. AHMT and KOH solutions were mixed from reagent reservoirs through the mixing column and reacted with formaldehyde getting through the hydrophobic porous polytetrafluoroethylene (PTFE) membrane to present unique colored products. The product color was finally collected and identified by a smart phone, showing a detection limit of 0.01 ppm. Yang et al. developed an array-assisted microfluidic chip based on functionalized SERS probes for aldehydes analysis [66]. The probe was served by Au@Ag nanocubs coated with MOF material [zeolite imidazolic acid skeleton-8 (ZIF-8)] and cysteamine (CA). Au@Ag nanocubs provided SERS enhancement, ZIF-8-protected core, and provided favorable reaction environment, and CA captured aldehyde gases. With the assistance of the array, trace and multicomponent detection was realized with the detection limit of 1 ppb.

4.2 Food safety

Food quality and safety is the primary issue to ensure human life and health, and the monitoring of food quality and safety has increasingly attracted people's attention. Due to the advantages of small equipment size, high throughput and low cost, microfluidic technology is widely used in the detection of pesticide residues, antibiotic residues, and additives in food.

4.2.1 Pesticide residues

Pesticides are powerful tools to protect and improve crop yields, but pesticide residues on food can also cause serious harm to human health. Traditional detection methods for pesticide residues, including enzyme inhibition, spectroscopy, and chromatography, can provide highly sensitive determination of a variety of pesticide residues, but these methods have disadvantages such as large equipment, high cost, and low degree of automation. Therefore, portable, low-cost, and automated pesticide residue microfluidic chips are gradually emerging.

Fernandez-Ramos et al. developed a bioactive μPAD based on enzyme inhibition method for colorimetric determination of organophosphate pesticides (OPs) and carbamate pesticides in water [67]. The μPAD consisted of a sampling zone at the bottom, an acetylcholine esterase (AChE) and acetylcholine chloride (AChCl) deposition zone separated in the middle, and a reaction zone containing bromocresol purple (BCP) as the pH indicator immobilized by sol-gel at the top. At the reaction zone, purple indicated the presence of pesticides, while yellow negated it. Pesticide quantification was achieved using the camera and Image J software, with detection limits of 0.24 and 2.00 μg/L for carbaryl and chlorpyrifos and reproducibility of 4.2–5.5%. Hu et al. developed a sensitive fluorescence microfluidic sensor based on quantum dot (QDs) aerogel and AChE for OP detection [68]. The red fluorescence of L-glutathione (GSH)-stabilized CdTe QDs aerogel was quenched by the reaction of AChE-catalyzed hydrolysis of acetylthiocholine (ATCh), but was recovered by the inhibition of OPs on the hydrolysis of ATCh. The researchers evaluated paraoxon, parathion, and dichlorvos, respectively, with a detection limit of 0.38 pM, and detection ranges from 10^{-5} to 10^{-12} M, demonstrating the high sensitivity and wide detection range of the QDs-AChE aerogel-based microfluidic arrays sensor. The OPs in fruit samples represented by apples were also detected with a recovery reaching 98%. Lafuente et al. combined SERS, polyoxometalate-decorated gold nanostructures (Au@POM), and microfluidic techniques for the analysis of organophosphorous pesticide paraoxon-methyl [69]. Au@POM (POM: $H_3PW_{12}O_{40}$ (PW) and $H_3PMo_{12}O_{40}$ (PMo)) were self-assembled in situ on the surface of PDMS microfluidic channel driven by UV-light without any extra functionalization procedure. POM provided high SERS enhancement factor by the charge transfer mechanism and excellent analyte capture due to the unique reactivity of metal-oxygen anion clusters. The SERS response of disposable Au@PW-based LOC was superior to that of re-used Au@PMo when evaluating the two kinds of Au@POM-based LOC using rhodamine R6G, but both showed an excellent analytical ability at the level of 10^{-6} M. The detection of OPs in dynamic conditions revealed the great application prospect of Au@PMo-based LOC in the real-time analysis of water pollutants. Yang et al. integrated enzyme inhibition and label-free screen-printed carbon electrode (SPCE) to develop a multilayer paper-based electrochemical microfluidic chip for pesticide residue identification [70]. The chip could identify abamectin, thioctyphos, and dimethoate by analyzing the impedance time-sequence spectrum data caused by enzyme inhibition during 15 minutes. The results of tests on actual samples (lettuce) with around 93% accuracy indicated a potential method for rapid, low-cost pesticide identification.

4.2.2 Antibiotic residues

Antibiotics are effective bacteriostatic agents used to treat diseases in humans and animals, but their overuse in recent years has led to fears of highly resistant superbugs. In response to the threat of antibiotics, the World Health Organization (WHO) has set maximum residual limits of antibiotics in food, triggering an explosion of research topics such as accurate identification and rapid detection of antibiotics.

Taghizadeh-Behbahani et al. developed a petal-shaped paper-based microfluidic based on a colorimetric sensor array for antibiotic residue detection in milk and eggs [71]. The determination of the photoelectric tongue device was based on the principle that different interactions of antibiotics with five metal indicator complexes produced specific color changes, combined with chemometrics data analysis for the data collection and processing of the collected images. In the real sample determination with external addition, the results of antibiotics overlapped because of the complex matrix of milk and yolk samples, but the antibiotics in egg white could be 100% distinguishable. Zhao et al. developed an automated fluorescent microfluidic system based on a competitive immunoassay for simultaneous determination of chloramphenicol (CAP) in 10 milk samples [72]. A single assay could be completed in 20 minutes, which was six times faster than conventional enzyme-linked immunosorbent assay (ELISA), indicating a promising portable fluorescence detector for the point-of-care testing (POCT) of CAP. Nguyen et al. developed a SERS microfluidic chip based on the nanocomposite consisting of graphene oxide and gold nanoparticle for highly sensitive and specific detection of kanamycin (KANA) [73]. The nanocomposite material was used as SERS substrate, while the Texas Red connected to switchable Beacon aptamers was served as Raman dye. The chip was reusable, providing a linear range of 1 to 100 nM with a low detection limit of 0.75 nM. Aptamers also play an important role in the electrochemical impedance biosensor of ampicillin. Xu's group developed an integrated microfluidic chip for online multi-residue analysis of fluoroquinolones (FQs) and amantadine (ATD) in combination with triple quadrupole mass spectrometry (QQQ-MS) [74]. The microfluidic chip had six parallel microchannels, consisting of a micropillar-array-microfiltration unit and a microsolid phase extraction (micro-SPE) column packed with hydrophilic-lipophilic balanced (HLB) particles. After filtration, extraction, and elution, the sample solution was immediately electrosprayed for MS analysis. The microfluidic device was successfully used to simultaneously determine seven kinds of FQs and one kind of ATD from the chicken muscle. The mean recoveries were 85.2–122% with the relative standard deviation (RSD) were 5.6–20.3%. This automated device, which integrated sample loading, filtration, extraction and online elution, could be an optimal quantitative screening method for food safety due to the advantages of low reagent consumption, simplified sample pretreatment, high throughput, and online detection.

4.2.3 Food additives

In order to prevent corrosion and improve the taste and appearance of food, many food additives are added during food processing, but excessive food additives will cause food safety and health problems. Fortunately, there are a number of strategies available for onsite detection of food additives. Cyclamate is widely used as a sugar substitute because it is far sweeter than sucrose, but excessive intake can also lead to many diseases. Liu et al. constructed a microfluidic PMMA/paper detection (MPD) system based on diazotization reaction for the specific determination of cyclamate in food [75]. Under the condition of light shielding, cyclamate sample injected into the sample chamber of MPD was diazotized with $NaNO_2$ and HCl to produce cyclohexyl nitrite, which then reacted with the Bratton-Marshall reagent to produce a purplish red azo product. The reactant images were collected by a camera and wirelessly transmitted to a smartphone for cyclamate concentration analysis with an LOD of 20 ppm. For the detection of cyclamate in 10 commercial food products, the concentration deviation of the MPD system was less than 4.8% from that of conventional liquid chromatography/tandem mass spectrometry (LC-MS/MS). Liu et al. developed a μPAD platform based on the Janovsky reaction theory for the detection of benzoic acid which is known as a common preservative [76]. The Janovsky reaction between NaOH and dinitrobenzoic acid obtained from the nitrification of benzoic acid was induced on a microheater, and the color changes were collected and analyzed by CMOS cameras and smartphones. The deviation between the results of benzoic acid in 21 commercial samples and that of standard HPLC was less than 6.6%. Other detection methods such as micro-point discharge emission spectroscopy (μPD-OES), SERS, and electrochemical amperometry can also be used to determine food additives [77–79].

4.3 Biomedicine

In recent years, the emergence of a large number of new viruses and bacteria, causing some very difficult diseases, has urged people to devote a lot of energy into the diagnosis and treatment of diseases. Among various detection methods, biosensing technology combined with microfluidics plays an important role in the early diagnosis of diseases, leading to a series of different application scenarios of POCT.

4.3.1 Infectious disease

In recent years, the spread of some infectious diseases with rapid onset, lack of effective drugs, and rapid transmission has seriously threatened the life safety of all mankind. Rapid and accurate detection of related pathogens can effectively prevent infectious diseases. Microfluidic technology shows great potential in this field. Huang et al. developed a microfluidic chip-based PCR array system called Onestart, which could complete the simultaneous detection of 21 kinds of respiratory pathogens, including coronavirus, influenza virus, and adenovirus [80]. The Onestart system was a fully integrated system where the sample cracking, nucleic acid extraction/purification and real-time fluorescent quantitative PCR processes were all performed on the same chip, ensuring that the entire analysis process took less than 1.5 hours. The Onestart system had a lower limit of detection of 1.0×10^3 copies mL^{-1} and 100% differentiation of reference sample mixtures, and showed consistency with real-time PCR analysis of batched clinical samples. This fast, sensitive, and specific detection Onestart system would help in clinical management of respiratory infections and prevention of pathogen transmission. Garg et al. developed an acoustic-driven ELISA microfluidic device for rapid detection of HIV, HPV, and HSV antibodies in serum and saliva to assess a variety of infectious diseases [81]. In this device, the flow and mixing of reagents were achieved with lateral cavity acoustic transducers (LCATs) and vertical cavity acoustic transducers (VCATs), respectively. The resulting convection would accelerate the binding of antigen and antibody and shorten the detection time. The device reduced the minimum time required to test serum samples from HIV, HPV, HSV positive patients, and normal donors from 2 hours for benchtop testing to 17.5 minutes. This technology could be extended to the rapid detection of protein biomarkers for a variety of infectious diseases. In addition, malaria virus [22, 82], Zika virus [83], rotavirus [84], etc., can be ultra-sensitive and rapidly detected by microfluidic devices.

4.3.2 Cancer

Cancer is the second leading cause of death worldwide, and its prevalence is increasing dramatically. Advanced detection technology is of great significance for early diagnosis of cancer, tracking of progress and evaluation of treatment. Microfluidic technology can detect a variety of cancer diagnostic factors from biological fluids.

Circulating tumor cells (CTC) carry important features of cancer progression and metastasis, providing important information for the detection of cancer. Chen et al. designed a field-effect transistor (FET)-based PDMS microfluidic chip for automatic detection and counting of CTC [85]. The chip used CTC-specific aptamers fixed on the top of the FET to identify CTCS in the blood and completed the efficient capture and count of cancer cells through the cell capture device, which could count up to 42 cancer cells. In addition to aptamers, peptides [86] and proteins [87, 88] are also commonly used for CTC capture and counting. Cell-free circulating tumor DNA (ctDNA), which is more stable and more abundant than CTCS, is also an important indicator for evaluating cancer. Using the separation effect of dimethyl dithiobispropionimidate (DTBP) on ctDNA, Jin et al. demonstrated a low-cost analytical system for detecting ctDNA in the plasma of patients with colorectal cancer [89]. The platform did not use dissociative reagents which would lead to ctDNA damage and increased cellular DNA background. Instead, DTBP that can bind to the amine group of ctDNA was used as the elution buffer and sodium bicarbonate was used as the separation agent. The separation of ctDNA could be completed in 15 minutes. Combining DTBP microfluidic platform with other detection methods (biooptical sensors, Sanger sequencing, and PCR methods), 71.4% of BRAF and KRAS mutational spectra in stage I-IV patients with colorectal cancer was detected. Genetic mutations in pancreatic, ovarian, and non-small cell lung cancers can also be detected using ctDNA [90, 91].

4.4 Other applications

In addition to analytical detection, microfluidic technology also has good uses in other fields. In chemical synthesis, microfluidic technology has excellent performance in the preparation of a variety of nanomaterials (including nanoparticles [92–95], quantum dots [96–98], colloids [99, 100], etc.). In medicine, microfluidic can not only be used for disease diagnosis but also play a great advantage in drug screening [101, 102], drug transportation [103–105], and so on. In addition, microfluidic technology promotes the development of some human organs-on-chip (OOC) [106–108], which is of great significance for human research on the physiological function of organs in normal and diseased states.

5. Conclusion

In recent years, microfluidic technology has developed rapidly, and more and more unit components are integrated in chips, and the integration scale is becoming larger and larger. At the same time, the microflow control chip can parallel process a large number of samples, with high throughput, fast analysis speed, low material consumption, and small pollution characteristics, providing a powerful platform for the foundation and application of environment, food safety, biomedicine, chemistry, and other fields.

While there are so many exciting and compelling developments, we still face challenges in the field of microfluidics, mainly in terms of the shift between theoretical research ideas and practical techniques for solving real-world problems. Therefore, in the future, we should still focus on basic research to promote the development of this field, but at the same time, we should also pay attention to the application of microfluidic technology, especially in the field of high-throughput applications. It is believed that microflow control products will soon enter the market one after another and play an important role in the healthy development of human life and the protection of ecological environment.

Acknowledgements

This work was funded by the National Key Research and Development Program of China (grant number 2023YFD1702104, 2021YFD2000204, and 2023YFD1701800); National Natural Science Foundation of China (grant number 32301688); and Science and Technology Mission Program of Anhui Province (grant number S2022t06010123).

Conflict of interest

The authors declare no conflict of interest.

Author details

Qiao Cao and Xiangyu Chen*
Intelligent Agriculture Engineering Laboratory of Anhui Province, Institute of Intelligent Machines, Hefei Institutes of Physical Science, Chinese Academy of Sciences, Hefei, China

*Address all correspondence to: cxy0910@iim.ac.cn

References

[1] Manz A, Graber N, Widmer HM. Miniaturized total chemical analysis systems: A novel concept for chemical sensing. Sensors and Actuators B: Chemical. 1990;**1**(1):244-248. DOI: 10.1016/0925-4005(90)80209-I

[2] Daw R, Finkelstein J. Lab on a chip. Nature. 2006;**442**(7101):367-367. DOI: 10.1038/442367a

[3] Janasek D, Franzke J, Manz A. Scaling and the design of miniaturized chemical-analysis systems. Nature. 2006;**442**(7101):374-380. DOI: 10.1038/nature05059

[4] Whitesides GM. The origins and the future of microfluidics. Nature. 2006;**442**(7101):368-373. DOI: 10.1038/nature05058

[5] Sackmann EK, Fulton AL, Beebe DJ. The present and future role of microfluidics in biomedical research. Nature. 2014;**507**(7491):181-189. DOI: 10.1038/nature13118

[6] Kou S, Cheng D, Sun F, et al. Microfluidics and microbial engineering. Lab on a Chip. 2016;**16**(3):432-446. DOI: 10.1039/C5LC01039J

[7] Zhang J, Yan S, Yuan D, et al. Fundamentals and applications of inertial microfluidics: A review. Lab on a Chip. 2016;**16**(1):10-34. DOI: 10.1039/C5LC01159K

[8] Martynova L, Locascio LE, Gaitan M, et al. Fabrication of plastic microfluid channels by imprinting methods. Analytical Chemistry. 1997;**69**(23):4783-4789. DOI: 10.1021/ac970558y

[9] McDonald JC, Duffy DC, Anderson JR, et al. Fabrication of microfluidic systems in poly(dimethylsiloxane). Electrophoresis. 2000;**21**(1):27-40. DOI: 10.1002/(SICI)1522-2683(20000101)21:1<27::AID-ELPS27>3.0.CO; 2-C

[10] Martinez AW, Phillips ST, Butte MJ, et al. Patterned paper as a platform for inexpensive, low-volume, portable bioassays. Angewandte Chemie, International Edition. 2007;**46**(8):1318-1320. DOI: 10.1002/anie.200603817

[11] Ge Q, Li Z, Wang Z, et al. Projection micro stereolithography based 3D printing and its applications. International Journal of Extreme Manufacturing. 2020;**2**(2):022004. DOI: 10.1088/2631-7990/ab8d9a

[12] Han W, Kong L, Xu M. Advances in selective laser sintering of polymers. International Journal of Extreme Manufacturing. 2022;**4**(4):042002. DOI: 10.1088/2631-7990/ac9096

[13] Sun Q, Xue Z, Chen Y, et al. Modulation of the thermal transport of micro-structured materials from 3D printing. International Journal of Extreme Manufacturing. 2022;**4**(1):015001. DOI: 10.1088/2631-7990/ac38b9

[14] Minas GMH, Catarino SO. Lab-on-a-chip devices for chemical analysis. In: Li D, editor. Encyclopedia of Microfluidics and Nanofluidics. Boston, MA: Springer US; 2013. pp. 1-22. DOI: 10.1007/978-3-642-27758-0_774-2

[15] Samiei E, Tabrizian M, Hoorfar M. A review of digital microfluidics as portable platforms for lab-on a-chip applications. Lab on a Chip. 2016;**16**(13):2376-2396. DOI: 10.1039/C6LC00387G

[16] Shi H, Zhao Y, Liu Z. Numerical investigation of the secondary flow effect of lateral structure of micromixing channel on laminar flow. Sensors and Actuators B: Chemical. 2020;**321**:128503. DOI: 10.1016/j.snb.2020.128503

[17] Basiri A, Heidari A, Nadi MF, et al. Microfluidic devices for detection of RNA viruses. Reviews in Medical Virology. 2021;**31**(1):e2154. DOI: 10.1002/rmv.2154

[18] Fukuba T, Fujii T. Lab-on-a-chip technology for in situ combined observations in oceanography. Lab on a Chip. 2021;**21**(1):55-74. DOI: 10.1039/D0LC00871K

[19] Sasaki R, Inagawa A, Xie X, et al. Absorption-based colorimetric detection of nickel(II) ion by phase separation of thermoresponsive magnetic nanoparticles under microflow. Analytical Sciences. 2024;**40**(4):791-798. DOI: 10.1007/s44211-024-00521-x

[20] Shen L, Hagen JA, Papautsky I. Point-of-care colorimetric detection with a smartphone. Lab on a Chip. 2012;**12**(21):4240-4243. DOI: 10.1039/C2LC40741H

[21] Kong M, Li Z, Wu J, et al. A wearable microfluidic device for rapid detection of HIV-1 DNA using recombinase polymerase amplification. Talanta. 2019;**205**:120155. DOI: 10.1016/j.talanta.2019.120155

[22] Xu G, Nolder D, Reboud J, et al. Paper-origami-based multiplexed malaria diagnostics from whole blood. Angewandte Chemie, International Edition. 2016;**55**(49):15250-15253. DOI: 10.1002/anie.201606060

[23] Liang L, Lan F, Yin X, et al. Metal-enhanced fluorescence/visual bimodal platform for multiplexed ultrasensitive detection of microRNA with reusable paper analytical devices. Biosensors and Bioelectronics. 2017;**95**:181-188. DOI: 10.1016/j.bios.2017.04.027

[24] Liu Z, Yang X, Zhang Y, et al. Hollow fiber SPR sensor available for microfluidic chip. Sensors and Actuators B: Chemical. 2018;**265**:211-216. DOI: 10.1016/j.snb.2018.03.030

[25] Huang C, Bonroy K, Reekmans G, et al. Localized surface plasmon resonance biosensor integrated with microfluidic chip. Biomedical Microdevices. 2009;**11**(4):893-901. DOI: 10.1007/s10544-009-9306-8

[26] Yang Z, Xia L, Li S, et al. Highly sensitive refractive index detection based on compact HSC-SPR structure in a microfluidic chip. Sensors and Actuators, A: Physical. 2019;**297**:111558. DOI: 10.1016/j.sna.2019.111558

[27] Zhang W, Li N, Lin L, et al. Concentrating single cells in picoliter droplets for phospholipid profiling on a microfluidic system. Small. 2020;**16**(9):1903402. DOI: 10.1002/smll.201903402

[28] Zhang Y, Li H, Ma Y, et al. Paper spray mass spectrometry-based method for analysis of droplets in a gravity-driven microfluidic chip. The Analyst. 2014;**139**(5):1023-1029. DOI: 10.1039/c3an01769a

[29] Nie L, Xu G-B, Wang X-Y, et al. Coupling microchip electrophoresis with MALDI-TOF-MS based on a freezing technique. Chinese Chemical Letters. 2013;**24**(6):491-493. DOI: 10.1016/j.cclet.2013.03.053

[30] Wang X-N, Song Y, Tang W, et al. Integration of fluorescence and MALDI imaging for microfluidic chip-based screening of potential

thrombin inhibitors from natural products. Biosensors and Bioelectronics. 2023;**237**:115527. DOI: 10.1016/j.bios.2023.115527

[31] Karikalan N, Yamuna A, Lee TY. Highly scalable design of 3D copper-doped graphene printed microfluidic chip for the electrochemical quantification of plasma homocysteine. Chemical Engineering Journal. 2023;**472**:145014. DOI: 10.1016/j.cej.2023.145014

[32] Park YM, Park J, Lim SY, et al. Integrated pumpless microfluidic chip for the detection of foodborne pathogens by polymerase chain reaction and electrochemical analysis. Sensors and Actuators B: Chemical. 2021;**329**:129130. DOI: 10.1016/j.snb.2020.129130

[33] Cao Y, Floehr J, Azarkh D, et al. Microfluidic aspiration-assisted electrical impedance spectroscopy system is a reliable tool for the characterization of oocyte hardening. Sensors and Actuators B: Chemical. 2023;**380**:133316. DOI: 10.1016/j.snb.2023.133316

[34] Cao Y, Floehr J, Ingebrandt S, et al. Dry film resist laminated microfluidic system for electrical impedance measurements. Micromachines. 2021;**12**(6):632. DOI: 10.3390/mi12060632

[35] Yiannacou K, Sariola V. Controlled manipulation and active sorting of particles inside microfluidic chips using bulk acoustic waves and machine learning. Langmuir. 2021;**37**(14):4192-4199. DOI: 10.1021/acs.langmuir.1c00063

[36] Liu X, Chen X, Yang Z, et al. Surface acoustic wave based microfluidic devices for biological applications. Sensors & Diagnostics. 2023;**2**(3):507-528. DOI: 10.1039/D2SD00203E

[37] Li J, Jamieson WD, Dimitriou P, et al. Building programmable multicompartment artificial cells incorporating remotely activated protein channels using microfluidics and acoustic levitation. Nature Communications. 2022;**13**(1):4125. DOI: 10.1038/s41467-022-31898-w

[38] Jeger-Madiot N, Mousset X, Dupuis C, et al. Controlling the force and the position of acoustic traps with a tunable acoustofluidic chip: Application to spheroid manipulations. Journal of the Acoustical Society of America. 2022;**151**(6):4165-4179. DOI: 10.1121/10.0011464

[39] Tajeddin A, Mustafaoglu N, Yapici MK, et al., editors. On-chip measurement of pH using a microcantilever: A biomimetic design approach. In: Symposium on Design, Test, Integration and Packaging of MEMS and MOEMS; 2021 August 25-27. Paris, France: IEEE; 2021. DOI: 10.1109/dtip54218.2021.9568499

[40] Chien C-C, Jiang J, Gong B, et al. AFM microfluidic cantilevers as weight sensors for live single cell mass measurements. Measurement Science and Technology. 2022;**33**(9):095009. DOI: 10.1088/1361-6501/ac7280

[41] Zhang R, Li Q, Tian L, et al. On-chip micro pressure sensor for microfluidic pressure monitoring. Journal of Micromechanics and Microengineering. 2021;**31**(5):055013. DOI: 10.1088/1361-6439/abf1b4

[42] Du Z, Zhao G, Wang A, et al. Pressure-triggered microfluidic contact lens for ocular drug delivery. ACS Applied Polymer Materials. 2022;**4**(10):7290-7299. DOI: 10.1021/acsapm.2c01118

[43] Yu N, Liu Y, Ji B, et al. High-sensitivity microliter blood pressure

sensors based on patterned micro-nanostructure arrays. Lab on a Chip. 2020;**20**(9):1554-1561. DOI: 10.1039/D0LC00063A

[44] Sunderman FW. Carcinogenic effects of metals. Federation Proceedings. 1978;**37**(1):40-46

[45] Tchounwou PB, Yedjou CG, Patlolla AK, et al. Heavy metal toxicity and the environment. In: Luch A, editor. Molecular, Clinical and Environmental Toxicology: Volume 3: Environmental Toxicology. Basel: Springer Basel; 2012. pp. 133-164. DOI: 10.1007/978-3-7643-8340-4_6

[46] Balali-Mood M, Naseri K, Tahergorabi Z, et al. Toxic mechanisms of five heavy metals: Mercury, lead, chromium, cadmium, and arsenic. Frontiers in Pharmacology. 2021;**12**:643972. DOI: 10.3389/fphar.2021.643972

[47] Wang G, Li J, Wu S, et al. A fully integrated, ready-to-use distance-based chemosensor for visual quantification of multiple heavy metal ions. Analytical Chemistry. 2022;**94**(46):15925-15929. DOI: 10.1021/acs.analchem.2c04712

[48] Huang W-H, Mai V-P, Wu R-Y, et al. A microfluidic aptamer-based sensor for detection of mercury(II) and lead(II) ions in water. Micromachines. 2021;**12**(11):1283. DOI: 10.3390/mi12111283

[49] Bai S, Serien D, Hu A, et al. 3D microfluidic surface-enhanced raman spectroscopy (SERS) chips fabricated by all-femtosecond-laser-processing for real-time sensing of toxic substances. Advanced Functional Materials. 2018;**28**(23):1706262. DOI: 10.1002/adfm.201706262

[50] Ma S, Zhao W, Zhang Q, et al. A portable microfluidic electrochemical sensing platform for rapid detection of hazardous metal Pb^{2+} based on thermocapillary convection using 3D Ag-rGO-f-$Ni(OH)_2$/NF as a signal amplifying element. Journal of Hazardous Materials. 2023;**448**:130923. DOI: 10.1016/j.jhazmat.2023.130923

[51] Lace A, Ryan D, Bowkett M, et al. Arsenic detection in water using microfluidic detection systems based on the leucomalachite green method. Analytical Methods. 2019;**11**(42):5431-5438. DOI: 10.1039/c9ay01580a

[52] Khanfar MF, Al-Faqheri W, Al-Halhouli AA. Low cost lab on chip for the colorimetric detection of nitrate in mineral water products. Sensors. 2017;**17**(10):2345. DOI: 10.3390/s17102345

[53] Vincent AG, Pascal RW, Beaton AD, et al. Nitrate drawdown during a shelf sea spring bloom revealed using a novel microfluidic in situ chemical sensor deployed within an autonomous underwater glider. Marine Chemistry. 2018;**205**:29-36. DOI: 10.1016/j.marchem.2018.07.005

[54] Ali MA, Jiao Y, Tabassum S, et al., editors. Electrochemical detection of nitrate ions in soil water using graphene foam modified by TiO_2 nanofibers and enzyme molecules. In: 2017 19th International Conference on Solid-State Sensors, Actuators and Microsystems (TRANSDUCERS), 2017 18-22 June. Kaohsiung, Taiwan: IEEE; 2017. DOI: 10.1109/TRANSDUCERS.2017.7994032

[55] Li X, Chang H. Chip-based ion chromatography (chip-IC) with a sensitive five-electrode conductivity detector for the simultaneous detection of multiple ions in drinking water. Microsystems & Nanoengineering.

2020;**6**(1):66. DOI: 10.1038/s41378-020-0175-x

[56] Ferreira FTSM, Mesquita RBR, Rangel AOSS. Novel microfluidic paper-based analytical devices (μPADs) for the determination of nitrate and nitrite in human saliva. Talanta. 2020;**219**:121183. DOI: 10.1016/j.talanta.2020.121183

[57] Fornells E, Murray E, Waheed S, et al. Integrated 3D printed heaters for microfluidic applications: Ammonium analysis within environmental water. Analytica Chimica Acta. 2020;**1098**:94-101. DOI: 10.1016/j.aca.2019.11.025

[58] Gallardo-Gonzalez J, Baraket A, Boudjaoui S, et al. A fully integrated passive microfluidic lab-on-a-chip for real-time electrochemical detection of ammonium: Sewage applications. Science of the Total Environment. 2019;**653**:1223-1230. DOI: 10.1016/j.scitotenv.2018.11.002

[59] Cho YB, Jeong SH, Chun H, et al. Selective colorimetric detection of dissolved ammonia in water via modified Berthelot's reaction on porous paper. Sensors and Actuators B: Chemical. 2018;**256**:167-175. DOI: 10.1016/j.snb.2017.10.069

[60] Clinton-Bailey GS, Grand MM, Beaton AD, et al. A lab-on-chip analyzer for in situ measurement of soluble reactive phosphate: Improved phosphate blue assay and application to fluvial monitoring. Environmental Science and Technology. 2017;**51**(17):9989-9995. DOI: 10.1021/acs.est.7b01581

[61] Luy EA, Morgan SC, Creelman JJ, et al. Inlaid microfluidic optics: Absorbance cells in clear devices applied to nitrite and phosphate detection. Journal of Micromechanics and Microengineering. 2020;**30**(9):095001. DOI: 10.1088/1361-6439/ab9202

[62] Grand MM, Clinton-Bailey GS, Beaton AD, et al. A lab-on-chip phosphate analyzer for long-term in situ monitoring at fixed observatories: Optimization and performance evaluation in estuarine and oligotrophic coastal waters. Frontiers in Marine Science. 2017;**4**:255. DOI: 10.3389/fmars.2017.00255

[63] Cheng YH, Barpaga D, Soltis JA, et al. Metal–organic framework-based microfluidic impedance sensor Platform for ultrasensitive detection of perfluorooctanesulfonate. ACS Applied Materials and Interfaces. 2020;**12**(9):10503-10514. DOI: 10.1021/acsami.9b22445

[64] Breshears LE, Mata-Robles S, Tang Y, et al. Rapid, sensitive detection of PFOA with smartphone-based flow rate analysis utilizing competitive molecular interactions during capillary action. Journal of Hazardous Materials. 2023;**446**:130699. DOI: 10.1016/j.jhazmat.2022.130699

[65] Guo X-L, Chen Y, Jiang H-L, et al. Smartphone-based microfluidic colorimetric sensor for gaseous formaldehyde determination with high sensitivity and selectivity. Sensors. 2018;**18**(9):3141. DOI: 10.3390/s18093141

[66] Yang K, Zong S, Zhang Y, et al. Array-assisted SERS microfluidic chips for highly sensitive and multiplex gas sensing. ACS Applied Materials and Interfaces. 2020;**12**(1):1395-1403. DOI: 10.1021/acsami.9b19358

[67] Fernández-Ramos MD, Ogunneye AL, Babarinde NAA, et al. Bioactive microfluidic paper device for pesticide determination in waters. Talanta. 2020;**218**:121108. DOI: 10.1016/j.talanta.2020.121108

[68] Hu T, Xu J, Ye Y, et al. Visual detection of mixed organophosphorous

pesticide using QD-AChE aerogel based microfluidic arrays sensor. Biosensors and Bioelectronics. 2019;**136**:112-117. DOI: 10.1016/j.bios.2019.04.036

[69] Lafuente M, Pellejero I, Clemente A, et al. In situ synthesis of SERS-active Au@POM nanostructures in a microfluidic device for real-time detection of water pollutants. ACS Applied Materials and Interfaces. 2020;**12**(32):36458-36467. DOI: 10.1021/acsami.0c06725

[70] Yang N, Zhou X, Yu D, et al. Pesticide residues identification by impedance time-sequence spectrum of enzyme inhibition on multilayer paper-based microfluidic chip. Journal of Food Process Engineering. 2020;**43**(12):e13544. DOI: 10.1111/jfpe.13544

[71] Taghizadeh-Behbahani M, Shamsipur M, Hemmateenejad B. Detection and discrimination of antibiotics in food samples using a microfluidic paper-based optical tongue. Talanta. 2022;**241**:123242. DOI: 10.1016/j.talanta.2022.123242

[72] Zhao M, Li X, Zhang Y, et al. Rapid quantitative detection of chloramphenicol in milk by microfluidic immunoassay. Food Chemistry. 2021;**339**:127857. DOI: 10.1016/j.foodchem.2020.127857

[73] Nguyen AH, Ma X, Park HG, et al. Low-blinking SERS substrate for switchable detection of kanamycin. Sensors and Actuators B: Chemical. 2019;**282**:765-773. DOI: 10.1016/j.snb.2018.11.037

[74] Rosati G, Ravarotto M, Scaramuzza M, et al. Silver nanoparticles inkjet-printed flexible biosensor for rapid label-free antibiotic detection in milk. Sensors and Actuators B: Chemical. 2019;**280**:280-289. DOI: 10.1016/j.snb.2018.09.084

[75] Liu C-C, Ko C-H, Fu L-M, et al. Light-shading reaction microfluidic PMMA/paper detection system for detection of cyclamate concentration in foods. Food Chemistry. 2023;**400**:134063. DOI: 10.1016/j.foodchem.2022.134063

[76] Liu C-C, Wang Y-N, Fu L-M, et al. Microfluidic paper-based chip platform for benzoic acid detection in food. Food Chemistry. 2018;**249**:162-167. DOI: 10.1016/j.foodchem.2018.01.004

[77] Kuek Lawrence CS, Tan SN, Floresca CZ. A "green" cellulose paper based glucose amperometric biosensor. Sensors and Actuators B: Chemical. 2014;**193**:536-541. DOI: 10.1016/j.snb.2013.11.054

[78] Zhu Y, Zhang L, Yang L. Designing of the functional paper-based surface-enhanced Raman spectroscopy substrates for colorants detection. Materials Research Bulletin. 2015;**63**:199-204. DOI: 10.1016/j.materresbull.2014.12.004

[79] Jing J, He S, Yang J, et al. Rapid and sensitive quantification of cyclamate in beverages by miniature microplasma optical emission spectrometry. Food Chemistry. 2023;**406**:135077. DOI: 10.1016/j.foodchem.2022.135077

[80] Huang E, Wang Y, Yang N, et al. A fully automated microfluidic PCR-array system for rapid detection of multiple respiratory tract infection pathogens. Analytical and Bioanalytical Chemistry. 2021;**413**(7):1787-1798. DOI: 10.1007/s00216-021-03171-4

[81] Garg N, Boyle D, Randall A, et al. Rapid immunodiagnostics of multiple viral infections in an acoustic microstreaming device with serum and saliva samples. Lab on a Chip. 2019;**19**(9):1524-1533. DOI: 10.1039/C8LC01303A

[82] Reboud J, Xu G, Garrett A, et al. Paper-based microfluidics for DNA diagnostics of malaria in low resource underserved rural communities. PNAS. 2019;**116**(11):4834-4842. DOI: 10.1073/pnas.1812296116

[83] Kaarj K, Akarapipad P, Yoon J-Y. Simpler, faster, and sensitive zika virus assay using smartphone detection of loop-mediated isothermal amplification on paper microfluidic chips. Scientific Reports. 2018;**8**(1):12438. DOI: 10.1038/s41598-018-30797-9

[84] Ye X, Xu J, Lu L, et al. Equipment-free nucleic acid extraction and amplification on a simple paper disc for point-of-care diagnosis of rotavirus A. Analytica Chimica Acta. 2018;**1018**:78-85. DOI: 10.1016/j.aca.2018.02.068

[85] Chen Y-H, Pulikkathodi AK, Ma Y-D, et al. A microfluidic platform integrated with field-effect transistors for enumeration of circulating tumor cells. Lab on a Chip. 2019;**19**(4):618-625. DOI: 10.1039/C8LC01072B

[86] Hassanzadeh-Barforoushi A, Warkiani ME, Gallego-Ortega D, et al. Capillary-assisted microfluidic biosensing platform captures single cell secretion dynamics in nanoliter compartments. Biosensors and Bioelectronics. 2020;**155**:112113. DOI: 10.1016/j.bios.2020.112113

[87] Kumeria T, Kurkuri MD, Diener KR, et al. Label-free reflectometric interference microchip biosensor based on nanoporous alumina for detection of circulating tumour cells. Biosensors and Bioelectronics. 2012;**35**(1):167-173. DOI: 10.1016/j.bios.2012.02.038

[88] Bravo K, Ortega FG, Messina GA, et al. Integrated bio-affinity nano-platform into a microfluidic immunosensor based on monoclonal bispecific trifunctional antibodies for the electrochemical determination of epithelial cancer biomarker. Clinica Chimica Acta. 2017;**464**:64-71. DOI: 10.1016/j.cca.2016.11.012

[89] Jin CE, Koo B, Lee TY, et al. Simple and low-cost sampling of cell-free nucleic acids from blood plasma for rapid and sensitive detection of circulating tumor DNA. Advanced Science. 2018;**5**(10):1800614. DOI: 10.1002/advs.201800614

[90] Gootenberg JS, Abudayyeh OO, Lee JW, et al. Nucleic acid detection with CRISPR-Cas13a/C2c2. Science. 2017;**356**(6336):438-442. DOI: 10.1126/science.aam9321

[91] Guan Y, Mayba O, Sandmann T, et al. High-throughput and sensitive quantification of circulating tumor DNA by microfluidic-based multiplex PCR and next-generation sequencing. The Journal of Molecular Diagnostics. 2017;**19**(6):921-932. DOI: 10.1016/j.jmoldx.2017.08.001

[92] Yagyu H, Tanabe Y, Takano S, et al. Continuous flow synthesis of monodisperse gold nanoparticles by liquid-phase reduction method on glass microfluidic device. Micro and Nano Letters. 2017;**12**(8):536-539. DOI: 10.1049/mnl.2017.0126

[93] Uson L, Arruebo M, Sebastian V, et al. Single phase microreactor for the continuous, high-temperature synthesis of <4 nm superparamagnetic iron oxide nanoparticles. Chemical Engineering Journal. 2018;**340**:66-72. DOI: 10.1016/j.cej.2017.12.024

[94] Gioria E, Signorini C, Wisniewski F, et al. Green synthesis of time-stable palladium nanoparticles using microfluidic devices. Journal of Environmental Chemical Engineering. 2020;**8**(5):104096. DOI: 10.1016/j.jece.2020.104096

[95] Xu L, Srinivasakannan C, Peng J, et al. Microfluidic reactor synthesis and photocatalytic behavior of Cu@Cu2O nanocomposite. Applied Surface Science. 2015;**331**:449-454. DOI: 10.1016/j.apsusc.2015.01.109

[96] Lignos I, Stavrakis S, Kilaj A, et al. Millisecond-timescale monitoring of PbS nanoparticle nucleation and growth using droplet-based microfluidics. Small. 2015;**11**(32):4009-4017. DOI: 10.1002/smll.201500119

[97] Tian Z-H, Xu J-H, Wang Y-J, et al. Microfluidic synthesis of monodispersed CdSe quantum dots nanocrystals by using mixed fatty amines as ligands. Chemical Engineering Journal. 2016;**285**:20-26. DOI: 10.1016/j.cej.2015.09.104

[98] Rao L, Tang Y, Li Z, et al. Efficient synthesis of highly fluorescent carbon dots by microreactor method and their application in Fe3+ ion detection. Materials Science and Engineering: C. 2017;**81**:213-223. DOI: 10.1016/j.msec.2017.07.046

[99] Yi G-R, Thorsen T, Manoharan VN, et al. Generation of uniform colloidal assemblies in soft microfluidic devices. Advanced Materials. 2003;**15**(15):1300-1304. DOI: 10.1002/adma.200304890

[100] Zhao LB, Pan L, Zhang K, et al. Generation of Janus alginate hydrogel particles with magnetic anisotropy for cell encapsulation. Lab on a Chip. 2009;**9**(20):2981-2986. DOI: 10.1039/B907478C

[101] Parent C, Raj Melayil K, Zhou Y, et al. Simple droplet microfluidics platform for drug screening on cancer spheroids. Lab on a Chip. 2023;**23**(24):5139-5150. DOI: 10.1039/D3LC00417A

[102] Zhai J, Liu Y, Ji W, et al. Drug screening on digital microfluidics for cancer precision medicine. Nature Communications. 2024;**15**(1):4363. DOI: 10.1038/s41467-024-48616-3

[103] Fabozzi A, Della Sala F, di Gennaro M, et al. Synthesis of hyaluronic acid core–shell nanoparticles via simple microfluidic-assisted nanoprecipitation method for active tumor targeting. New Journal of Chemistry. 2022;**46**(41):19763-19772. DOI: 10.1039/D2NJ03279A

[104] Hasani-Sadrabadi MM, Karimkhani V, Majedi FS, et al. Microfluidic-assisted self-assembly of complex dendritic polyethylene drug delivery nanocapsules. Advanced Materials. 2014;**26**(19):3118-3123. DOI: 10.1002/adma.201305753

[105] Han Z, Lv W, Li Y, et al. Improving tumor targeting of exosomal membrane-coated polymeric nanoparticles by conjugation with aptamers. ACS Applied Bio Materials. 2020;**3**(5):2666-2673. DOI: 10.1021/acsabm.0c00181

[106] Li LK, Huang W-C, Hsueh Y-Y, et al. Intramuscular delivery of neural crest stem cell spheroids enhances neuromuscular regeneration after denervation injury. Stem Cell Research & Therapy. 2022;**13**(1):205. DOI: 10.1186/s13287-022-02877-1

[107] Gerigk M, Bulstrode H, Shi HH, et al. On-chip perivascular niche supporting stemness of patient-derived glioma cells in a serum-free, flowable culture. Lab on a Chip. 2021;**21**(12):2343-2358. DOI: 10.1039/D1LC00271F

[108] Siwczak F, Loffet E, Kaminska M, et al. Intestinal stem cell-on-chip to study human host-microbiota interaction. Frontiers in Immunology. 2021;**12**:798552. DOI: 10.3389/fimmu.2021.798552

Section 3

Crystallization

Chapter 6

Impacts of Sulfur Curing Systems on Vulcanizations and Mechanical Performances of Elastomers: A Model Study Based on Sulfur Curing Systems and NR/SBR Blends

Ruogu Tang

Abstract

Vulcanization is one of the most significant procedures in elastomer product processing. The components and their proportions of the curing systems significantly determined the vulcanization processes and mechanical properties of the products. To understand this effect, a model study was adopted for investigation, with sulfur curing system as the model curing system and natural rubber/styrene-butadiene rubber blends as model elastomers. By adjusting the doses of sulfur and sulfur/accelerant ratio, the entire sulfur curing systems were divided into three classifications, i.e., conventional vulcanization systems (CV), effective vulcanization systems (EV), and semi-effective vulcanization systems (SEV). Under these divisions, NR/SBR vulcanizate blends were obtained via seven specific curing systems. Upon preparation, the NR/SBR vulcanizates were thoroughly tested for mechanical properties. In a general trend, CV-based vulcanizates showed the advantages of tensile and tear strengths, EV-based vulcanizates possessed higher Young's modulus and hardness, and SEV-based vulcanizates performed higher abrasion resistances. In addition, for each individual system, there would be an optimum sulfur/accelerant ratio by which the vulcanizates could be produced with enhanced mechanical properties.

Keywords: curing system, crosslink type, crosslink density, crystallization, orientation

1. Introduction

Vulcanization is a critical procedure in elastomeric product processing. During the vulcanization, the crosslinking formed between the polymer chains [1]. With appropriate specific curing systems, vulcanization could effectively improve the performances and stabilities of rubber products [1, 2]. Despite the development of novel curing systems in the lab during the past years, sulfur-based curing system is still one of the most widely used curing systems in the rubber industry [3]. Based on the sulfur usages and proportions, the sulfur curing systems could be categorized as conventional vulcanization systems (CV), effective vulcanization systems (EV) and semi-effective vulcanization systems (SEV) [3]. A major distinction among these classifications is the content and proportion of sulfur in the formula compositions, specially its ratio to vulcanization accelerator (abbreviated as sulfur/accelerant ratio). In CV systems, the sulfur/accelerant ratios are supposed to be greater than 2.5, while for EV systems, these ratios are less than 0.4. In the case of SEV systems, the sulfur/accelerant ratios are approximately around 1. It was estimated that during the vulcanization, the sulfur contents and the ratio to accelerants directly affect the formation of crosslinks, consequently affecting the configurations and/or conformations of chain segments, stereoregularities of polymer chains, and crystallization and/or orientation behaviors, thereby determining the mechanical properties of the vulcanizates [3]. Therefore, designing an appropriate curing system is a significant step for rubber processing. Some studies have indicated that the CV system is favorable for producing vulcanizates for dynamic applications, while the EV system should be considered for obtaining vulcanizates for static and hot environments [4–8]. However, these findings specifically focused on one type of curing system or a given example, which did not comparatively and comprehensively investigate the impacts induced by the chemical components and proportions (especially sulfur/accelerant ratios) of the curing systems, left some confusion and limitations in the applications of sulfur curing systems.

Therefore, to comprehensively understand the effects of sulfur curing systems on the vulcanization processes and related mechanical properties. A comprehensive set of vulcanizates were prepared through three sulfur curing systems. Natural rubber (NR) and styrene-butadiene rubber (SBR) were used as model elastomers due to their widespread usage. The NR and SBR were physically mixed and then vulcanized via CV, EV and SEV systems. After preparation, the vulcanizates were characterized by their mechanical properties, which included tensile strengths, tear strengths, elongations at break, Young's modulus, hardness (Shore A) and relative volume abrasions.

2. Model study

2.1 Experiment

In this study, NR, SBR and other chemicals were purchased from Qingdao Chemical Company, two roll rubber mill was obtained from Dongguan Cfine Machinery Co. Ltd., rotorless curemeter, tensile tester (electronic mode), Shore durometer and rotating cylindrical drum device were purchased from GoTech Co Ltd., densimeter was purchased from Preciso Co Ltd. The scanning electron microscope (SEM) was purchased from JEOL.

The Formula composition of raw rubbers (in powder state), sulfur and other additives for preparations of NR/SBR vulcanizates were listed in the following **Table 1**.

NR/SBR physical mixture compound based on the ISO 2393:2014. The mixtures were vulcanized with a standard protocol ISO 3417:1991. The curing temperature was maintained at 145°C and the pressure was kept at 10 MPa, respectively. The whole curing procedures were investigated based on ISO 6502-1:2018. Upon the

	CV	EV	SEV
NR	360	360	360
SBR	240	240	240
High abrasion furnace black (HAF)	150	150	150
Silica	100	100	100
Oil	100	100	100
Zinc Oxide	30	30	30
Stearic acid	12	12	12
Paraffin	6	6	6
N-cyclohexyl-2-benzothiazolesulphenamide (CBS vulcanizator)	Custom designed	Custom designed	Custom designed
A. isopropyl-N′-phenyl-P-phenylenediamine (4010NA, Antioxidant)	6	6	6
Tetramethylthiuram disulfide (TMTD, extra accelerator)	1.2	7.2	3.6
$CaCO_3$	30	30	30
Sulfur	22	4	8.4

Table 1.
Formula compositions of vulcanizates in each curing system (unit: G).

preparations, the vulcanizates were naturally cooled down. After that the samples were sputter coated and observed via scanning electron microscopy (SEM) to characterize the surface morphologies.

Mechanical properties were tested following the International Organization for Standardization (ISO). Specifically, tensile strengths were tested based on ISO 37:2017, tear strengths were tested based on ISO 34-1:2022, rebound resiliences were tested based on ISO 4662:2017, Young's modulus were tested based on ISO 1827:2016 and the data were recorded at 300% of elongation, elongations (at break points) were tested based on ISO 2285:2019, Shore A hardness were tested on ISO 48-4:2018, relative volume abrasions were tested based on the ISO 4649:2017. The physical mixture of NR/SBR was also tested for tensile strength as the control.

3. Results and discussion

In this study, different NR/SBR vulcanizates were prepared under three curing systems. Due to the variations of the components in each sample, the experiment settings were adjusted individually, and **Table 2** showed the optimum curing time for each sample obtained from the screening test. The NR/SBR vulcanizates were observed by SEM, as displayed in **Figure 1**. All the vulcanizates possessed a homogeneous phase, indicating the qualities that validate vulcanizates for mechanical tests. **Figure 2** showed the tensile stress/strain properties of the vulcanizates. Compared to the physical mixture of NR/SBR, the vulcanizates under each curing system presented higher tensile stress and modulus, confirming the enhancement effect of vulcanization on rubber samples. From the stress/strain curves, it could

	Test sample #	Sulfur/accelerant ratio
CV	1	3.75
	2	3.33
	3	2.92
SEV	4	1.14
EV	5	0.42
	6	0.36
	7	0.2

Table 2.
Designs of test samples.

also be found that the vulcanizates still present elasticities under initial elongation (**Table 3**) [9].

Mechanical properties of vulcanizates were shown from **Figures 3-9**. It could be found that the vulcanizates prepared by CV systems performed better in tensile and tear strengths (**Figures 3** and **4**). In addition, those prepared through EV systems had higher Young's modulus and Shore A hardness (**Figures 6** and **8**), while those products prepared through the SEV system presented higher abrasion resistance (**Figure 9**).

The impacts of sulfur curing systems could be attributed to their effects on the formations of crosslink bond crosslink densities. Based on the atoms engaged and the lengths of bonds, in sulfur curing systems, the crosslink bonds were generally divided into three types, i.e., polysulfur bond (mainly formed in CV systems), disulfide bond (usually formed in both EV and SEV systems) and monosulfur bonds (usually formed in SEV systems). The most significant differences among these types are the lengths of bonds and the bond energy. As shown in **Table 4**, the polysulfur bond has the lowest bond energy, bringing in the highest flexibility among all the candidates [9]. With such high flexibility, the polysulfur bonds are easy to reshape when encountering stress concentrations, and the polymer chains have enough time for orientation and crystallization. Besides, polysulfur bonds with low energy are susceptible to decomposition and rearrangement. All these effects contributed to diminishing stress concentrations. Meanwhile, with the absorbance of bond energy, the dynamic losses were reduced, accompanied by the rearrangement process. On the contrary, the disulfide and monosulfur bonds are usually short since only a few atoms were involved in forming the bonds. And high bond energies make them rigid. All these suggested that the bonds tend to decompose rapidly under stress, leaving the chains with insufficient time for orientation and crystallization. Also, both disulfide and monosulfur bonds are common chemically insert, making them less accessible to other bonds or radicals and rearrange once they are broken. The fragmented chains would result in uneven and demolished strengths. Therefore, vulcanizates equipped with high polysulfur bond contents provided better strengths in general conditions.

Young's modulus and Shore A hardness could be affected by the stress relaxation behaviors of polymer chains. The results showed that EV-based vulcanizates possessed better Young's modulus and Shore A hardness (**Figures 6** and **8**), and these would be ameliorated with the increment of the sulfur/accelerant ratio. This phenomenon could be related to the different stress relaxation behaviors in each

DOI: http://dx.doi.org/10.5772/intechopen.111727

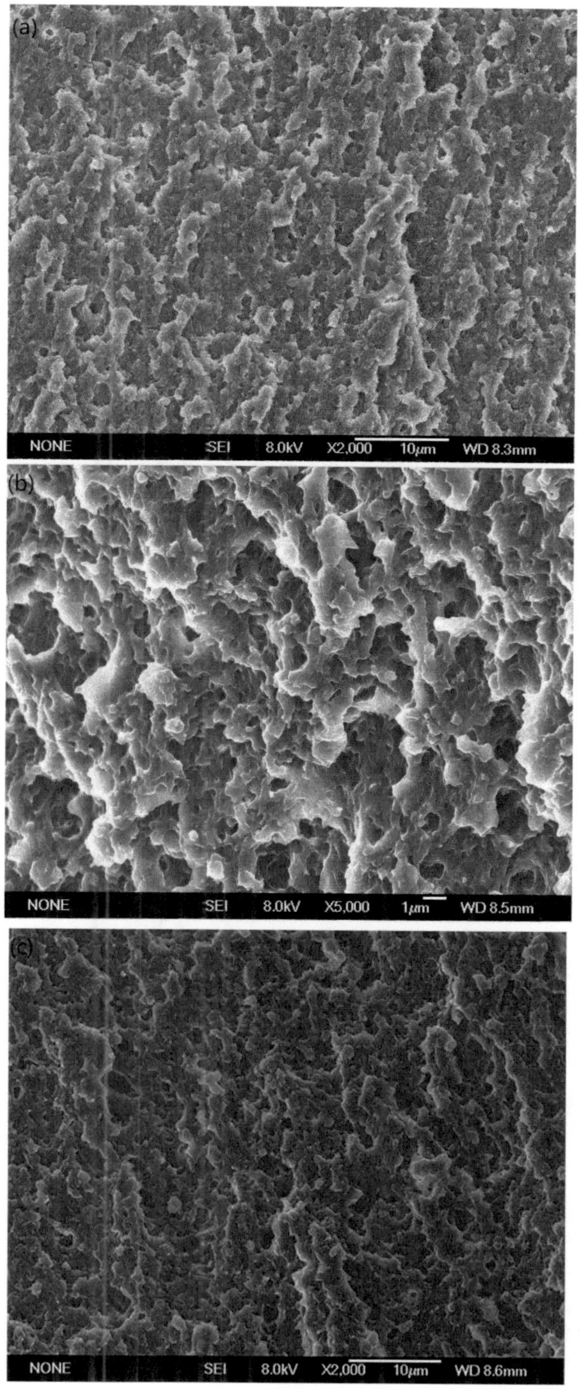

Figure 1.
SEM observations of NR/SBR vulcanizates. (a). From CV systems (sample 1); (b). From EV system (sample 6); (c). From SEV system (sample 4).

vulcanizate under different curing systems. The Parallel Maxwell Model (shown below) could be applied to illustrate the differences in stress relaxation rates in three curing systems [10].

Test sample #	Optimum curing time
1	6′44”
2	7′41”
3	6′15”
4	5′06”
5	7′00”
6	9′55”
7	10′18”

Table 3.
Curing time for each sulfur/accelerant ratio setting.

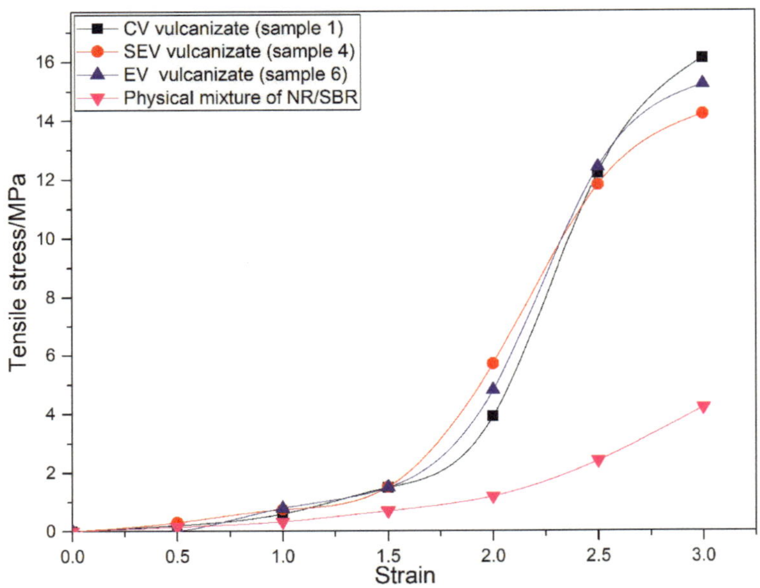

Figure 2.
Tensile stress/strain curves of vulcanizates under three curing systems.

$$\sigma = \sum \sigma_{0i} e^{\frac{-t}{\tau_i}} E = \sum \frac{\sigma_{0i}}{\varepsilon_0} e^{\frac{-t}{\tau_i}}$$

Specifically, σ_{0i} represents the stress of each independent Maxwell unit, τ_i represents the individual relaxation time of each unit, and E is the sum of the modulus as an entirety. As described before, polysulfur bonds usually have more chances to reshape and/or rearrange. Therefore, those vulcanizates composed of polysulfur bonds had faster stress relaxations than those dominated by disulfide bonds and monosulfur bonds. This means that in CV systems, τ_i could be remarkably shorter, and Young's modulus (as well as the hardness) could not be that high. In contrast, in EV and SEV systems, due to the reshape/rearrange-retardance effects caused by disulfide and monosulfur bonds, more time was needed for stress relaxations. Therefore, as the τ_i in each unit increases, Young's modulus and hardness would improve.

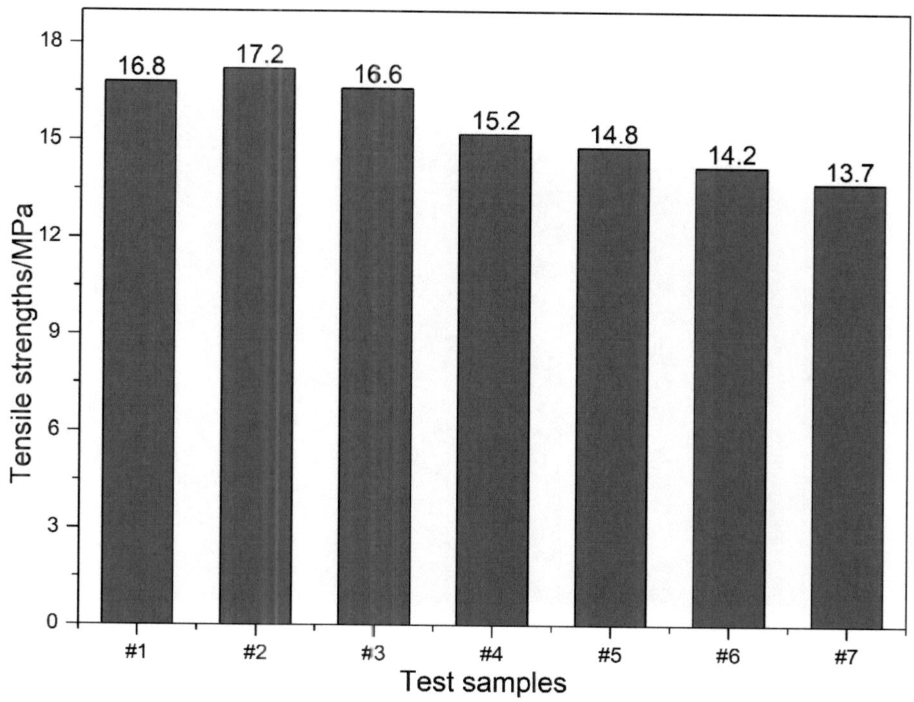

Figure 3.
Tensile strengths of NR/SBR vulcanizates obtained from different curing systems. (unit: MPa).

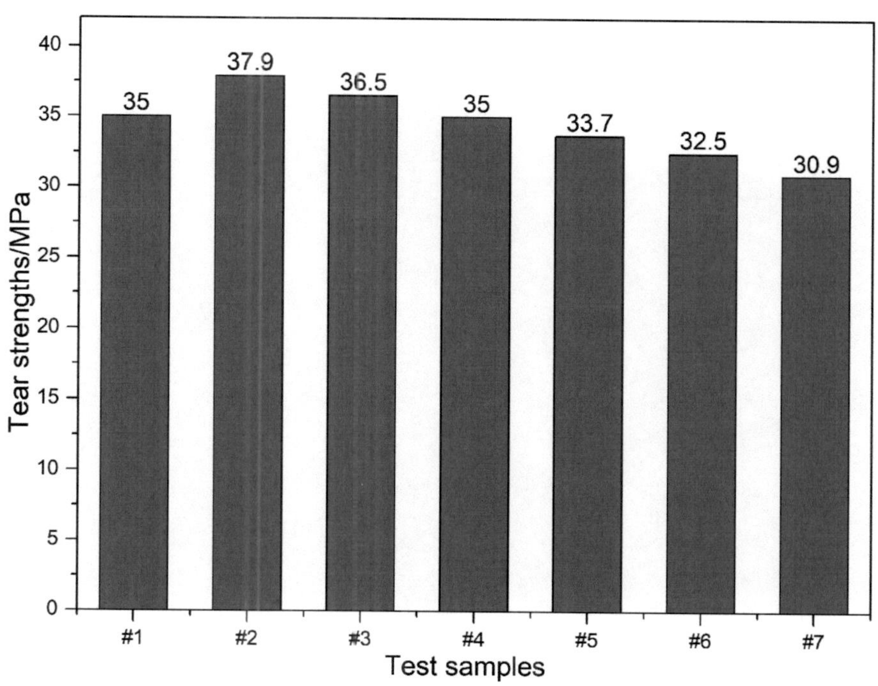

Figure 4.
Tear strengths of NR/SBR vulcanizates obtained from different curing systems. (unit: MPa).

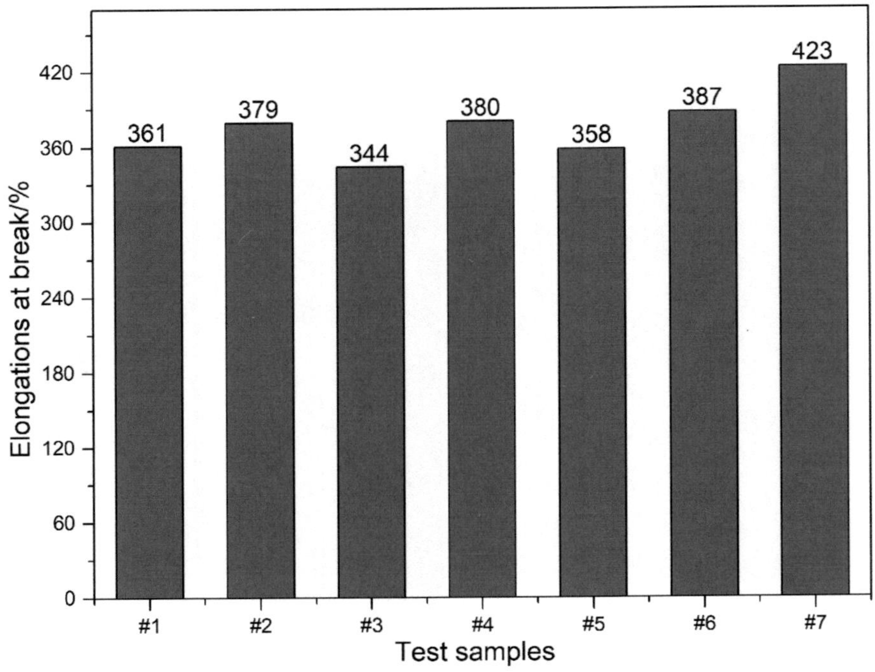

Figure 5.
Elongations at break of NR/SBR vulcanizates obtained from different curing systems. (unit: %).

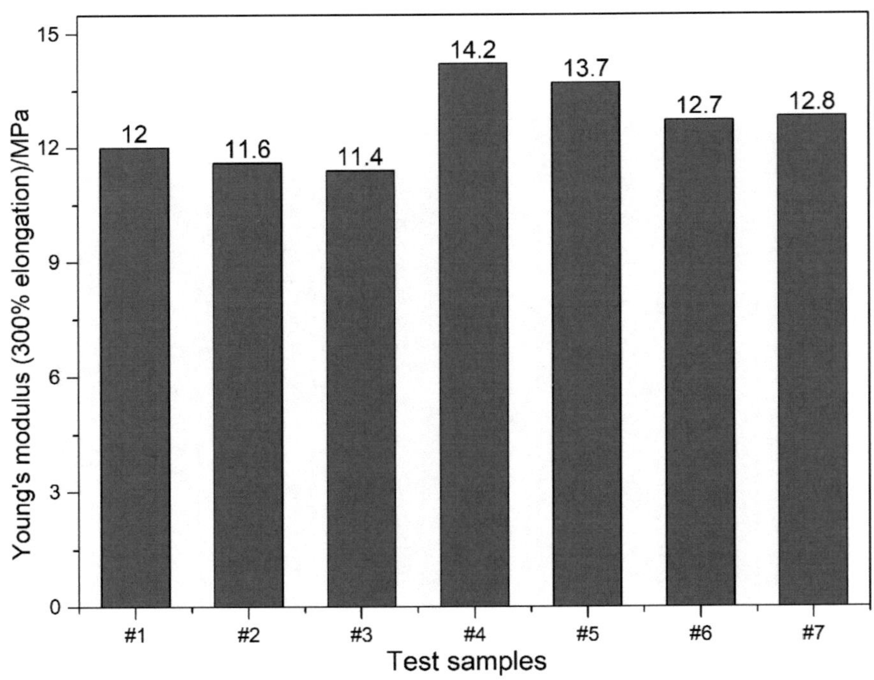

Figure 6.
Young's modulus of NR/SBR vulcanizates obtained from different curing systems. (unit: MPa).

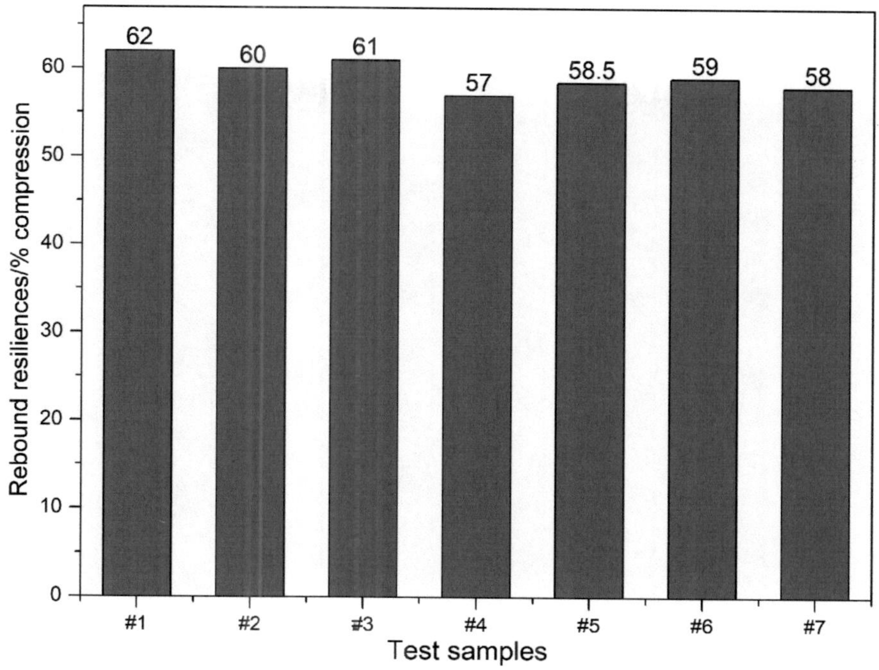

Figure 7.
Rebound resiliences of NR/SBR vulcanizates obtained from different curing systems. (unit: % compression).

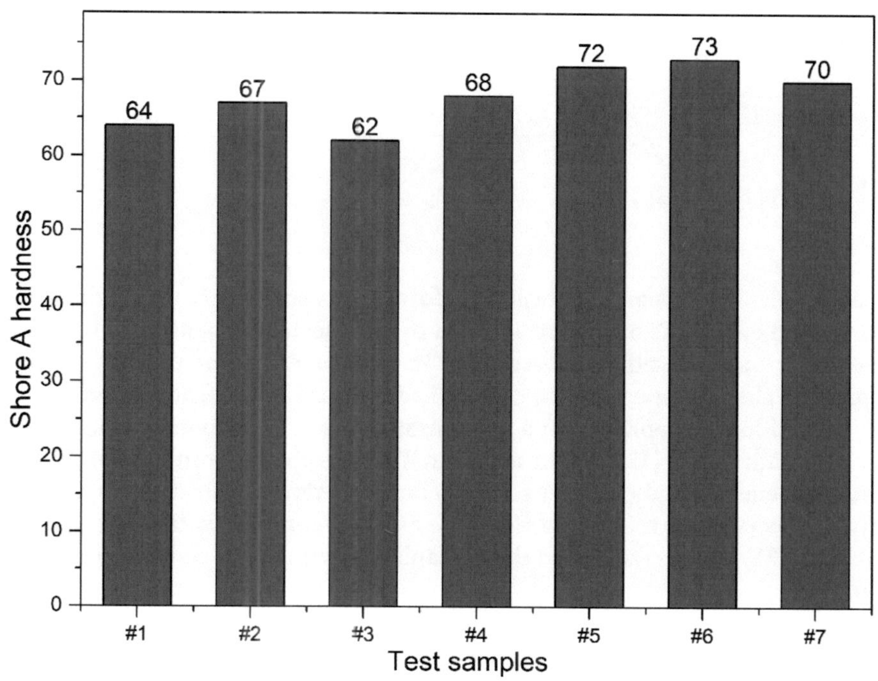

Figure 8.
Shore a hardness of NR/SBR vulcanizates obtained from different curing systems.

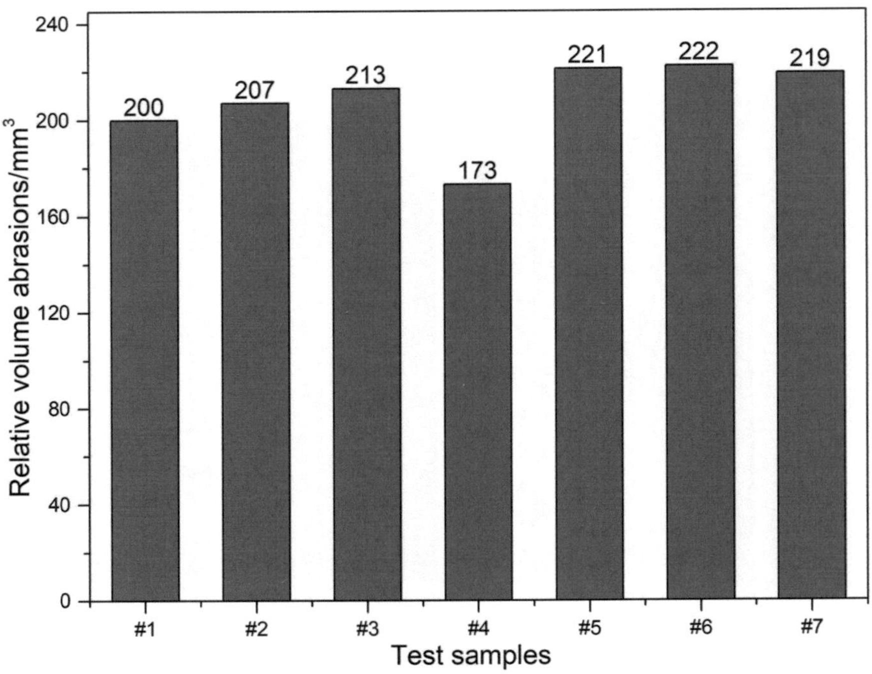

Figure 9.
Relative volume abrasions of NR/SBR vulcanizates obtained from different curing systems. (unit: mm^3).

Types of crosslink bond	Primary existence	Average bond energy/KJ·mol^{-1}
Monosulfur-based	In SEV system	295.70
Disulfide-based	In EV/SEV system	266.90
Polysulfur-based	In CV system	<264.50

Table 4.
Common types of sulfur crosslink bonds and corresponding bond energies.

The vulcanizates' dynamic abrasion performances, such as wear abrasions, fatigue abrasions, and curl abrasions, were affected by the mechano-chemical active chain-oxidized reactions induced by stresses [11]. In most oxidized and thermal oxidized reactions, the chains (especially the crosslinked parts) will break down and form ether bonds. In addition, the abrasion of a vulcanizate is directly proportional to its level of broken crosslink bonds [12]. As presented in **Table 4**, the monosulfur bond has the highest bond energy, and this keeps it from breaking down against stress. Consequently, it was inferred that under the same external conditions, the monosulfur bond-based SEV system facilitated the vulcanizates with enhanced abrasion resistances.

Figures 3–9 also proved the importance of the sulfur/accelerant ratio in any curing system by significantly determining the mechanical properties regardless of the other external factors. For example, considering a vulcanizate under CV systems, with the sulfur proportion (to accelerant) increasing, its tensile strengths, tear strengths, fractured elongations and Shore A hardness raised initially, promoted to the optimum and eventually retreated, the Young's modulus continuously improved, the relative volume abrasions continuously decreased, and the rebound resiliences fell at the beginning and then regained. As for vulcanizate under EV systems, inputting more sulfur is a positive approach to improve tensile and tear strengths, but this does not work well on other mechanical performances, as the elongations would reduce, the rebound resiliences and Relative volume abrasions stayed within a narrow variation, the Young's modulus as well as Shore A hardness varied randomly without regularities. By comparison, products from the SEV system presented an intermediate and moderate level of mechanical properties.

The sulfur/accelerant ratio of the vulcanizate is closely related to the its crosslink formations, including crosslink bonds distributions and densities, which further determines its mechanical behaviors, and these impacts are effective in either static or dynamic states. For example, with the augmentation of crosslink densities, polymer chains could efficiently react with each other, the structures of vulcanizates tend to become three-dimensional rather than linear, and chain segments need to overcome more restrictions to make movements [13]. Therefore, in general conditions, the Young's modulus, hardness, strengths and elongations improve with the increment of crosslink densities. However, when crosslink density reached the peak, these properties did not further enhance. This phenomenon can be explained by vulcanizates' crystallization and orientation characteristics.

The crystallization behaviors could be illustrated by Avrami formula: $X_{c,t}^{V} = X_{c,\infty}^{V}\left(1 - e^{-kt^{n}}\right)$, where X repents the degrees of crystallinities of vulcanizates, k represents the crystallization rate constant of a given polymer, t represents the total time to be crystallized and n served as Avrami index [10]. For all the curing systems in this study, no additional crystallization agents were used. Therefore, it could be assumed that the crystal nucleus formed in a homogeneous condition, and the formed structures were mainly oblique crystals. With this three dimensional uniformed nucleation, the n value can be fixed as 4 [13], so only t and k would influence the crystallinity level. k is theoretically constant, but when crosslink density exceeds the level that the nucleus could endure, excessive crosslink bonds will occupy the space initially used for crystallization. The remaining space may not be enough, so the crystallization becomes more challenging, and rate constant k should be considered smaller than before. Therefore, more time would be inevitably required for vulcanizates to crystallize with external stress. Under the same processing (curing) time, the degree of crystallinity would be lowered, consequently reducing the strengths.

The level of orientation of a polymer chain could be represented by the orientation factor formula: F = (3cos2θ-1)/2, where the θ represents the average angle between the direction of the vulcanizates' polymer chain and the direction of the orientation (0° to 180°) [12]. As the less crystallized polymers, the polymer chains of NR and SBR usually prefer to orient randomly. With external stress, chains will orient directly the same as direction of the stress, and the angle disappears (θ is counted as 0°), and F will be 1 (fully oriented). However, if the crosslinks are over-concentrated, the concentration parts will cause interruptions to the orientation even under external forces. The orientation direction will be altered and no longer directly identical to the stress, and there will be a new angle. If the θ is not 0°, the F will become lower than predicted. As a consequence, the strengths of vulcanizates would be damaged.

On the other side, the experimental settings, especially the curing time and temperature in this study, often serve a crucial role in the formation of crosslinks. Usually, the formation of crosslink bonds requires sufficient time within a complete curing process [14]. Given sufficient curing, the vulcanizates could have high crosslink densities. But if the vulcanizate underwent overdosed heating, chain breaking down could occur, during which the fragments randomly rearrange or react intramolecularly and intermolecularly. The unpredicted and uncontrolled reactions alter the appearance and qualities of vulcanizate products [15, 16]. Supported by Van't Hoff equation and Arrhenius equation, promoting the heating temperature will reduce the curing time, but overheating also possibly causes chain breakdown and damages the mechanical properties [17]. Therefore, in this study, during the curing process the temperatures for each sample were set below 150°C.

4. Conclusions

Based on the results and analysis from this model study, it was concluded that sulfur curing systems served as decisive factor in the preparation of vulcanizates, different types of sulfur curing systems provided the vulcanizates with various features and performances. In a general trend:

1. CV system can provide vulcanizates with relatively higher strengths.

2. EV system can provide vulcanizates with relatively better Young's modulus and Shore A hardness.

3. SEV can provide better abrasion resistances.

Another conclusion is that the mechanical properties of vulcanizates varied non-linearly according to variations of sulfur contents. Usually, an optimum sulfur/accelerant ratio could be confirmed by screening tests. Though more investigations are needed for further research, this model study could be used as guidance and for choosing the appropriate curing system for rubber products processing.

Acknowledgements

We gratefully appreciate the technical support and financial assistance from the aforementioned suppliers.

Conflict of interest

The authors declare no conflict or competition of interests.

Abbreviations

NR	Natural rubber
SBR	Styrene-butadiene rubber
CV	Conventional vulcanization
EV	Effective vulcanization
SEV	Semi-effective vulcanization

DOI: http://dx.doi.org/10.5772/intechopen.111727

Author details

Ruogu Tang
Qingdao University of Science and Technology, Qingdao, China

*Address all correspondence to: ruogutang@gmail.com

References

[1] Zhang Z, Guo F, Ke Y, Xiang C, Jia X. Effect of vulcanization on deformation behavior of rubber seals: Thermal–mechanical–chemical coupling model, numerical studies, and experimental validation. Materials & Design. 2022; **224**:111314. DOI: 10.1016/j.matdes.2022.111314

[2] Albuini-Oliveira NM, Rubinger M, Rabello AS, Vidigal AEC, Visconte LY, Lopes TC, et al. The influence of ammonium and phosphonium salts on natural rubber vulcanization with experimental and commercial accelerators. Polymer Bulletin. 2023;**80**: 3717-3743. DOI: 10.1007/s00289-022-04236-9

[3] Andrea D, Rigotti D, Fredi G. Recent advances in the devulcanization technologies of industrially relevant sulfur-vulcanized elastomers. Advanced Industrial and Engineering Polymer Research. 2022;**10**:1-22. DOI: 10.1016/j.aiepr. 25 2022.11.003

[4] Sun B, Li J, Xiang L, Lin F, Che L, Tian W, et al. Simulating vulcanization process during tire production to explore sulfur migration during pyrolysis. Fuel. 2022;**330**:125665. DOI: 10.1016/j.fuel.2022.125665

[5] Nardelli F, Calucci L, Carignani E, Silvia B, Cettolin M, Arimondi M, et al. Influence of sulfur-curing conditions on the dynamics and crosslinking of rubber networks: A time-domain NMR study. Polymers. 2022;**14**(4):767. DOI: 10.3390/polym14040767

[6] Soares BG, de Oliveira M, Meireles D, Sirqueira AS, Mauler RS. Dynamically vulcanized polypropylene/nitrile rubber blends: The effect of peroxide/bis-maleimide curing system and different compatibilizing systems. Journal of Applied Polymer Science. 2008;**110**(6):3566-3573. DOI: 10.1002/app.28946

[7] Maciejewska M, Baranowska AS. The synergistic effect of dibenzyldithiocarbamate based accelerator on the vulcanization and performance of the silica-filled styrene-butadiene elastomer. Materials. 2022;**15**(4):1450. DOI: 10.3390/ma1504 1450

[8] Wei Y, Liu G, Zhang H, Zhao F, Luo M, Liao S. Non-rubber components tuning mechanical properties of natural rubber from vulcanization kinetics. Polymer. 2019;**183**:121911. DOI: 10.1016/j.polymer.2019.121911

[9] El-Nemr KF. Effect of different curing systems on the mechanical and physico-chemical properties of acrylonitrile butadiene rubber vulcanizates. Materials & Design. 2011; **32**(6):3361-3369. DOI: 10.1016/j.matdes.2011.02.010

[10] Sukcharoen K, Noraphaiphipaksa N, Hasap A, Kanchanomai C. Experimental and numerical evaluations of localized stress relaxation for vulcanized rubber. Polymers. 2022;**14**(5):873. DOI: 10.3390/polym14050873

[11] Zhu L, Xu L, Jie S, Li B. Preparation of styrene-butadiene rubber vitrimers with high strength and toughness through imine and hydrogen bonds. Industrial & Engineering Chemistry Research. 2023;**62**(5):2299-2308. DOI: 10.1016/10.1021/acs.iecr.2c03133

[12] Khiêm VN, Le-Cam JB, Charlès S, Itskov M. Thermodynamics of strain-induced crystallization in filled natural rubber under uni- and biaxial loadings, part II: Physically-based constitutive

theory. Journal of the Mechanics and Physics of Solids. 2022;**159**(2):104712. DOI: 10.1016/j.jmps.2021.104712

[13] Gedde Ulf W, Hedenqvist MS. Fundamental Polymer Science. 2nd ed. Cham, Switzerland: Springer; 2019. p. 189. DOI: 10.1007/978-3-030-29794-7

[14] Liu H, Wang x, Jia D. Recycling of waste rubber powder by mechano-chemical modification. Journal of Cleaner Production. 2020;**245**(1):118716. DOI: 10.1016/j.jclepro.2019.118716

[15] Polgar LM, Kingma A, Roelfs M, van Essen M, van Duin M, Picchioni F. Kinetics of cross-linking and de-cross-linking of EPM rubber with thermoreversible Diels-Alder chemistry. European Polymer Journal. 2017;**90**: 150-161. DOI: 10.1016/j.eurpolymj. 2017.03.020

[16] Loos K, Aydogdu AB, Lion A, Johlitz M, Calipel J. Strain-induced crystallisation in natural rubber: A thermodynamically consistent model of the material behaviour using a multiphase approach. Continuum Mechanics and Thermodynamics. 2020; **32**:501-526. DOI: 10.1007/s00161-019-00859-y

[17] Mousavi MM, Hosseinnezhad S, kabir SF, Burnett DJ, Fini EH. Reaction pathways for surface activated rubber particles. Resources, Conservation and Recycling. 2019;**149**:292-300. DOI: 10.1016/j.resconrec.2019.05.041